Cracking
THE
GOLDEN STATE
EXAMINATION

Chemistry

The Princeton Review

Cracking

THE GOLDEN STATE EXAMINATION

Chemistry

by Amanda Stewart

Random House, Inc.
New York
www.randomhouse.com/princetonreview

Princeton Review, L.L.C.
2315 Broadway
New York, NY 10024

E-mail: comments@review.com

Copyright © 2000 by Princeton Review Publishing, L.L.C.

All rights reserved under International and Pan-American Copyright Conventions.

Published in the United States by Random House, Inc., New York

ISBN 0-375-75357-5

Editor: Rachel Warren
Production Editor: Kristen Azzara
Production Coordinator: Scott Harris

Manufactured in the United States of America.

9 8 7 6 5 4 3 2 1

First Edition

ACKNOWLEDGMENTS

I would like to thank Mom & Dad for all their support and guidance, Rick Sliter for having confidence in me, and Rachel Warren for her undying patience. I would also like to thank Laurie Barnett, Kristen Azzara, and Scott Harris for putting it all together.

Contents

CHAPTER 1: THE MYSTERY EXAM:
ABOUT THE GOLDEN STATE EXAMINATIONS 1

CHAPTER 2: STRUCTURE AND STRATEGIES 9

CHAPTER 3: ATOMIC STRUCTURE AND THE PERIODIC TABLE 17

CHAPTER 4: BONDING 33

CHAPTER 5: STOICHIOMETRY 47

CHAPTER 6: OXIDATION-REDUCTION REACTIONS 57

CHAPTER 7: GASES AND PHASE PHANGES 65

CHAPTER 8: SOLUTIONS 79

CHAPTER 9: KINETICS AND EQUILIBRIUM 85

CHAPTER 10: ACIDS AND BASES 95

CHAPTER 11: THERMODYNAMICS 109

CHAPTER 12: ELECTROCHEMISTRY 119

CHAPTER 13: NUCLEAR DECAY 127

CHAPTER 14: ORGANIC CHEMISTRY AND BIOCHEMISTRY 135

CHAPTER 15: LABORATORY 145

CHAPTER 16: PRACTICE TEST ONE **155**
ANSWERS AND EXPLANATIONS 167

CHAPTER 17: PRACTICE TEST TWO **175**
ANSWERS AND EXPLANATIONS 185

CHAPTER 18: PRACTICE TEST THREE **195**
ANSWERS AND EXPLANATIONS 207

CHAPTER 19: PRACTICE TEST FOUR **215**
ANSWERS AND EXPLANATIONS 225
About the Author 235

Chapter 1

THE MYSTERY EXAM: ABOUT THE GOLDEN STATE EXAMINATIONS

WHAT ARE THE GOLDEN STATE EXAMS?

Golden State Examinations (GSEs) were established by the State of California Board of Education in 1983 (So your parents never took them—they've probably even never heard of them and may not understand how important they are.). The tests are designed to offer a rigorous examination in key academic subjects to students in grades 7 to 12. Students who pass have a variety of advantages over those who don't, including the fact that their school transcripts will be more attractive to college admissions boards.

The GSE program has grown in the last few years, both in the number of different exams offered and the number of students who take at least one GSE. During the 1999–2000 academic year, thirteen different GSE exams will be administered, and California students will complete over one million examinations. In 1998, more than 2,100 graduates earned the Golden State Seal Merit Diploma, which recognizes students who have mastered their high school curriculum (see below for more information on the Merit Diploma).

You probably purchased this book because a teacher recommended it or because you were told to get ready to take a GSE. Well, lucky for you, this is the most complete guide available to prepare you for the GSEs. Our goal is simple—to get you ready for the GSE in Chemistry.

WE'RE HERE TO HELP

The GSEs are somewhat of a mystery to students. There are no official practice tests that you can study. Our research and development team has spent countless hours to ensure that this guide tells you everything you need to know to ace the GSEs. All of the information that's released about the GSEs is included in this book. We've also spoken to students and teachers about their experiences with these tests and have designed our content review around their feedback. In short, we have the inside scoop on the GSEs and we're going to share it with you.

What's so Special About This Book?

GSEs test subject knowledge as well as the application of that knowledge. The goal of this book is two-fold. First, we want to help you remember or relearn some of the subject material that's covered in the exam. Second, we want you to become familiar with the structure of the GSE so that you'll know exactly what to do on test day.

We at The Princeton Review aren't big fans of standardized tests, and we understand the stress and challenge that a GSE presents. But with our techniques and some work on your part, you should be able to do well on these tests. There's just one more thing you might be wondering—why *should* I take the GSEs?

WHY SHOULD I TAKE THE GOLDEN STATE EXAMINATIONS?

There are many reasons why you should spend time and energy getting ready for the GSEs. They include:

- **Qualification for a Golden State Seal Merit Diploma**

    The Golden State Seal Merit Diploma is one of the—if not *the*—highest academic awards given by the State of California. In the pages that follow, we will detail exactly how you can qualify for a Merit Diploma.

- **Recognition on Your Transcript**

    If you perform well on a Golden State Examination, you will receive recognition on your high school transcript. We'll tell you how the scoring system and award processes work in the pages that follow.

- **College Admissions Committees Will Love You**

    A strong performance on the GSEs will make you look great to colleges and universities; they're academic awards that will demonstrate to schools that you can excel in an academic environment.

- **They Are Risk Free**

    First, there is absolutely no fee to take a Golden State Examination, so you won't need to worry about spending money on these tests. Second, there is absolutely no penalty if you do not perform well on a Golden State Examination. A score that does not give you an award will not appear on your transcript. In fact, only you will know if you did *not* pass a Golden State Examination. So you've got nothing to lose and a lot to gain!

- **All the Things Your Teacher Would Say**

    There are academic benefits to passing the tests as well. If you asked your teacher, "Why should I take these exams?" your teacher would probably tell you that, in addition to all the benefits listed above, These tests provide a great opportunity for you to demonstrate what you have learned throughout high school, with the possibility of receiving numerous awards and titles for strong academic performance. The GSEs are a challenge that can enrich your high school experience.

    Although we'll be a little less formal in the way we say it, we agree with the teacher's advice. The GSEs are your chance to show off what you know. You should receive recognition for all your hard work and you should have all the necessary tools to ensure that you do well.

HOW ARE STUDENTS RECOGNIZED FOR THEIR PERFORMANCE?

If you score within the highest levels on any one GSE, you will receive one of three awards: high honors, honors, or recognition. Say you take three GSEs—one in chemistry, one in biology, and one in written composition. Because you used The Princeton Review test-prep books, you can give yourself a pat on the back. You are among the one-third range of all students who pass the GSEs, earning the awards listed below:

Test	Award
Written Composition	High Honors
Chemistry	Honors
Biology	Recognition

These awards are formally called Academic Excellence Awards. This means that students who receive one of these three awards will receive an Academic Excellence Award from the State of California. This will be recorded on your high school transcript and you'll receive a gold insignia on your diploma if you get high honors or honors.

Here is a better description of what the levels of honors on the GSE mean:

- **High Honors:** This is the most prestigious award given to students on the Chemistry GSE. It will be given to about the top 10 percent of students. If you receive "high honors" on the Chemistry GSE, you will receive a special gold seal on your high school diploma and the award will be placed on your transcript. Further, you can use this result as part of the requirements necessary for pursuing the ultimate award, the Golden State Seal Merit Diploma.

- **Honors:** This is the second most prestigious award given to students on the Chemistry GSE. It will be given to about 12 percent of the students that take the exam (students who score in the 78th to 90th percentile). If you receive "honors" on the Chemistry GSE, you will receive the same rewards as students who achieved a score of "high honors."

- **Recognition:** This is the final type of Academic Excellence award given to students on the Chemistry GSE. It will be given to approximately 15 percent of the students who take the exam (students who score in the 66th to 78th percentiles). If you get a "recognition" on the GSE in Chemistry, you will receive notification of this achievement on your high school transcript. Further, you can use this as part of the requirements necessary for pursuing the ultimate award, the Golden State Seal Merit Diploma.

Any one of these awards signals high achievement to colleges, universities, and employers. Finally, Golden State scholars are eligible to pursue a Golden State Seal Merit Diploma.

What is the Golden State Seal Merit Diploma?

In July of 1996, the State of California developed a Golden State Seal Merit Diploma program to recognize high school graduates who demonstrated high performance in several different academics areas. The Golden State Seal Merit Diploma is the most prestigious award you can receive by taking the GSEs. In 1997, the first year that the Golden State Seal Merit Diploma was issued, more than 1,300 high school seniors received the award. This number jumped to more than 2,100 in 1998, and will continue to increase as more students take the GSEs.

To receive the Merit Diploma, students must receive high honors, honors, or recognition designations on *six* different GSEs. The specific tests and requirements are described below.

How Can I Get a Golden State Seal Merit Diploma?

You do not need to apply to receive a Golden State Seal Merit Diploma. You just need to complete four required examinations, plus two elective exams, and receive at least recognition for them. School districts track the performance of each student and submit the information to the California Department of Education.

The four exams that students must pass are:

1. English (Written Composition or Reading and Literature)

2. U.S. History

3. Mathematics (Algebra, Geometry, or High School Mathematics)

4. Science (Biology, Chemistry, Physics, or Coordinated Science)

In addition to the four required exams, you may take two other GSEs selected from the following: Economics, Spanish Language, or Government and Civics. You may also complete an *additional* science, mathematics, or English exam as one of your two electives. For example, let's say you complete both the Chemistry and Geometry examinations and receive Academic Excellence awards on each exam. In this case, one of them will be counted as the mathematics *required* exam; the other will be counted as an elective.

What Material is Covered on the GSE?

GSEs are developed by a committee of teachers, university professors, and other education specialists. Each examination is designed and tested so that the content reflects the state standards for each subject. In general, you should expect that the information on a GSE will be similar to what you've been tested on during the academic year. The style and format of the GSE may be different, but the material should be just like the stuff you studied in class. Unlike many other high school examinations the GSEs are not designed to trick or trap you.

Most examinations consist of one day of multiple-choice questions, followed by one day of written work. In the case of Chemistry, you will perform a laboratory. All GSEs consist of two 45-minute parts. In Chapter 2, we'll discuss the specific format, structure, and scoring of the Chemistry Golden State Examination.

So Here's the Deal.

Listed below are all the frequently asked questions about the administration of the GSEs. If there's anything we don't cover or if you're still confused, ask your guidance counselor or teacher.

How are GSE Exams Scored?

Every Golden State Examination has specific scoring criteria based on its format, structure, and level of difficulty. See Chapter 2 for more specific information about how the Chemistry GSE is scored.

As we mentioned, there is no penalty whatsoever for poor performance on a Golden State Examination. If you fail to receive one of the honor designations we discussed earlier, there will be no mention of it on your academic transcript. Further, students who do not receive an honors designation on one GSE should still be encouraged to take additional GSEs. Each test is scored independently. Performance on one GSE will have no impact on the scoring of any other GSE.

Remember only about one-third of all test takers are honored for their performance on the Golden State Examination. These tests aren't a shoo-in. You need to know the material and be familiar with the structure to pass. This book is your key to being among the well prepared.

> For additional information about the GSE program, contact the Standards, Curriculum, and Assessment Division of the California Department of Education:
>
> Phone: (916) 657-3011
>
> Fax: (916) 657-4964
>
> E-mail: star@cde.ca.gov
>
> Internet: www.cde.ca.gov/cilbranch/sca/gse/gse.html

When Will I Receive My GSE Results?

Results from the GSEs are first sent to your school district. If you take a winter examination, you should expect to hear about your results in May. If you take a spring examination, you should expect to hear about your results once you return to school in the fall.

If you have any questions regarding your performance on the GSE, you can talk to your high school counselor for some more information about this.

Can I Take the Test More Than Once?
No. Students are eligible to take each GSE only one time. For this reason, be sure that you are prepared to take each Golden State Examination.

How Can I Keep Track of All the GSE Tests and Requirements?
Determining which GSEs to take, and when to take them, can be a confusing process. The California Department of Education has designed some worksheets for use by students that will help you keep track of this information. Ask your high school counselor for a copy of these worksheets.

How Do I Inform Colleges About My Golden State Awards?
If you are applying to a college, university, or military academy, you will want to make sure that any awards you've received on the GSEs are included in your application. If you received high honors, honors, or recognition on a GSE, this will be noted on your high school transcript. In addition, you can get a form called the *GSE Status Report for College Applications*. This form is available from your high school counselor and, along with your high school transcript, it will ensure that admissions boards notice your performance on the tests.

HOW THIS BOOK IS ORGANIZED
The next chapter of this book is devoted to giving you the specifics about the test you're about to take. We will discuss test structure, format, and scoring, and we'll also talk about some techniques and strategies that can be helpful to you. Our goal is to provide you with a "bag of tricks" that you can use throughout this exam.

We will then provide a specific content review of the subject material that's covered on the GSE. Rather than giving you lists of things to know, our goal is to give you information so that you can use it on the GSE. How do we know what is tested on the GSEs? We have carefully studied California State Curriculum Standards and Golden State Examination questions, surveyed high school teachers, and reviewed textbooks to determine exactly what is covered on each test. We'll use sample questions throughout the review to show you how certain topics are tested on the GSE.

Finally, we have prepared and constructed four full-length practice tests for the Golden State Examination. We will provide you with detailed explanations to each problem, and sample written work, when appropriate. Use these tests to recognize the areas in which you need improvement.

WHAT IS THE PRINCETON REVIEW?

The Princeton Review is the nation's leader in test preparation. We have offices in more than fifty cities across the country, as well as many outside the United States. The Princeton Review supports more than two million students every year with our courses, books, on-line services, and software programs. In addition to helping high school students prepare for the GSEs, we help them with the SAT-I, SAT-II, PSAT, and ACT, along with many other statewide standardized tests. The Princeton Review's strategies and techniques are unique and, most of all, successful.

Remember, this book will work best in combination with the material you have learned throughout your high school course. Our goals are to help you remember what you have been taught over the past year, and show you how to apply this knowledge to the specific format and structure of the Golden State Examination.

AND FINALLY...

We applaud your efforts to spend the time and energy to prepare for the Golden State Examination in Chemistry. You are giving yourself the opportunity to be rewarded for your academic achievement. Remember that the GSE will not test you on information you have never seen before. A strong year in Chemistry, combined with a review of the material and the test-taking strategies in this book, will leave you more than prepared to handle the GSE. Don't become frustrated if you don't remember everything at once; it may take some time for the information and skills to come back. Stay focused, practice, and try to have fun working through this book. Good luck!

Chapter 2

STRUCTURE AND STRATEGIES

It may seem pretty intimidating that only one-third of all students who take the GSEs receives any sort of honors. You might be wondering whether you can be one of them . . .but of course you can! Just remember that most students don't prepare at all before taking the GSEs, so you're already ahead of the game.

In this chapter, we'll tell you exactly which chemistry concepts are tested, and how. We'll also give you an idea of the scoring process, the structure, and the format of the test. We'll also show you what you'll need to do to receive an Academic Excellence Award (high honors, honors, or recognition). It's crucial that you use our techniques when you're taking the test. We'll refer to these techniques throughout the book to make sure you incorporate them into your practice.

WHAT IS TESTED ON THE CHEMISTRY GSE?

The content of the Chemistry GSE is in alignment with the State Board *California Chemistry Academic Content Standards*. This means that what's tested on the Chemistry GSE will be similar to the information you were presented during the academic year. Specifically, the following content areas are covered:

- **Basic foundations**

    Atomic and molecular theory; notations, formulas and models; stoichiometry, equations, mole concept; Periodic table (organizations, trends, and applications); physical properties

- **States of Matter**

    Properties of solids, liquids, and gases; changes of state, energy changes; gas laws (relationships and applications); solutions

- **Bonding**

    Principles of ionic, covalent (polar and nonpolar), and metallic bonding; attractions among particles (polar and nonpolar interactions, relationship to physical and chemical properties)

- **Reactions**

    Reactions in aqueous solution (precipitation, acid-base, oxidation-reduction); equilibrium and rates of reaction; energy changes during reactions; practical applications

In the chapters ahead, we will provide substantial review in each of the topic areas that will be covered on the Chemistry GSE.

HOW IS THE CHEMISTRY GSE EXAM STRUCTURED?

The GSE in Chemistry is in two parts, administered in two 45-minute sections. For example, Part I may be on a Tuesday, and Part II on a Wednesday.

Part I consists of approximately thirty multiple-choice questions and one written-response question. In general, these questions are designed to test a wide range of chemistry concepts. You will probably find several of these questions easy, but you may find that you are unfamiliar with the concepts tested in others. Don't worry, we'll review all the material you need to know. Each of the questions consists of four answer choices. Later, we'll talk about how to use the answer choices to your advantage. The written response question will have a couple of parts and ask you to answer them completely and with obvious hard knowledge of the topic.

Part II consists of a laboratory section. The laboratory sections will be performed independently by each student at a station. You will be required to use regular laboratory equipment and chemicals and operate using the usual safety procedures. The literature published about this exams states that the laboratory section also requires you to record your observations and data in an accurate and organized way, support all your analyses with calculations and specific evidence, and use scientific arguments to substantiate your methods.

HOW IS THE TEST SCORED?

A machine will score all of Part I. Chemistry teachers and other professionals will score the written response and laboratory portion of the Chemistry exam. Your performance on Part I, combined with your performance on Part II of the Chemistry GSE will determine your overall score.

CAN I USE A CALCULATOR ON THIS TEST?

Yes! A calculator may be required for some of the problems on the Chemistry GSE. You should use the calculator that you have used throughout your math classes. You can use any calculator you want, except for ones with QWERTY (typewriter) keyboards.

WHERE CAN I FIND A REAL GSE EXAM?

Sample copies of the real GSEs are not available, but you should get enough practice from the four full-length diagnostic Chemistry GSEs in the back of this book, which are followed by explanations. These tests simulate the format and kinds of questions you can expect to see on the GSE.

YOUR BAG OF TRICKS

Have you ever seen the cartoon *Felix the Cat*? Felix fought crime, solved problems, and got his way out of difficult situations by reaching into his bag of tricks. In this special bag, he'd find the exact tool he needed to resolve any situation. With his bag of tricks, Felix was invincible.

In this section, we'll help you fill up your own bag of tricks. What will be in there? Strategies and tools for handling each type of question on the Chemistry GSE, as well as general strategies for how to approach the test. It is important to know that being a smart test-taker is just as important as knowing the material tested. Managing your time, knowing when to guess, and knowing what the questions are *really* asking are skills you can learn, and we'll teach them to you. As you'll see, there is a difference between knowing the material, and being able to apply it to the test.

Let's take an example of two students, Gretchen and Laurie, each with the same amount of chemistry knowledge. Now, Gretchen took the same math class as Laurie, but Gretchen has received additional training. She has learned to think like the people who write the GSE; she understands their traps, knows fast ways to eliminate incorrect answer choices, and has the best techniques to use for certain types of problems. In short, she has learned how to become a solid test-taker. Gretchen, with her bag of tricks, is now going to do much better on the GSE than Laurie. Why? Not because she knows more, but because she knows how to take this specific test in a smarter way than Laurie does. She understands the rules of the game. Once you know the rules of the game, you know how best to apply your skills to the game.

General Strategies

Now that you know what is tested on the Chemistry GSE, and in what format it is tested, we need to talk about the best way to take this test. In the pages that follow, we'll discuss some general tools for you to use as you proceed through the test. Starting in Chapter 3, we will discuss specific strategies for particular *types* of questions.

An Empty Scantron Sheet is a Bad Scantron Sheet

In the past, you've probably taken a standardized test that had a guessing penalty. This penalty meant that points would be subtracted from your raw score if you answered a question incorrectly. Guessing penalties are meant to discourage test takers from answering every question. But guess what? There is *no* guessing penalty on the GSE! Your score is only determined by the number of questions that you get correct; it doesn't matter how many questions you get incorrect. So when you take the GSE, there is one thing that you must do before you turn in your test: You must answer every single multiple-choice question. There are thirty questions on Part I of the Chemistry GSE. Before you turn in your test, make sure that you have selected an answer for all thirty. Earning an Academic Excellence Award may boil down to just one additional point, and leaving a question blank guarantees a wrong answer.

So now you know that you must choose an answer for every question. Great, now let's talk about how to be an intelligent guesser.

Process of Elimination (POE)

Try the following question:

What is the capital of Malawi?

Unsure? Do you know even where Malawi is located? If not, don't panic. Geography and world capitals are not topics tested on the GSE (especially the Chemistry test!). If you had to answer this question without any answer choices, you'd probably be in trouble. You'd just randomly pick a city, and most likely guess wrong.

Of course, on the GSE, you will have answer choices to choose from. Rather than closing your eyes and selecting an answer at random, take a look at the choices—you might find some information that can help you.

What is the capital of Malawi?

A. Paris

B. Lilongwe

C. New York

D. London

Now do you know? Can you identify any answer choices that you know are *not* correct? Well, you can probably eliminate A, C, and D. Although you probably didn't know that Lilongwe was the capital of Malawi, you could tell that it was the correct answer by eliminating incorrect answer choices. This procedure is called Process of Elimination, or POE for short.

Process of Elimination will help you become a better guesser. This is because oftentimes, it's easier to see incorrect answer choices than it is to pinpoint the correct one. Remember to *cross out* any answer choice that you know is incorrect, then, if you still need to make a guess, select an answer from your remaining choices.

It is rare that POE will actually help you eliminate three answer choices like we did in the sample problem above. However, every time you get rid of one answer choice, the odds of getting that question correct go up significantly. Instead of a 25 percent chance of guessing correctly, you might find yourself guessing with a 50 percent (1 in 2) or 33 percent (1 in 3) chance of getting a question right.

Let's try another example that would be more likely to appear on the Chemistry GSE:

1. $2NO_2(g) \leftrightarrow N_2O_4(g)$

    If pressure is increased to the above equilibrium expression, keeping temperature constant, which of the following will occur?

A. The reaction will shift to the left

B. The reaction will shift to the right

C. There will be no effect

D. The volume will also increase

We'll review exactly how to answer equations that involve LeChatelier's principle, but for now, let's think about the situation. You probably know that when pressure is increased in a reaction system, *something* happens to the reaction, shifting it away from equilibrium. This means that you can eliminate choice C. Now, take a look at the answer choices. D says that the volume will increase if the pressure is increased. Does this make sense? No, in fact the opposite it true, so eliminate choice D. At this point, if you can't remember that increasing the pressure of the system shifts the reaction in the direction in which fewer moles of gas are produced, you could guess and still have a 50% chance of answering the question correctly. Of course, if you did remember that then you're home free. The answer is choice B.

Process of Elimination is such an important concept that we'll be referring to it throughout this book, including in the explanations provided for the practice tests. Some specific POE strategies will apply to certain chemistry questions that will be presented in the chapters ahead. It is important that you practice using POE, because getting rid of incorrect answers is a powerful tool on the GSE.

You Don't Have to Start at Number 1

Some tests contain an order of difficulty within each section. On these kinds of tests, the first question is generally very easy, and the questions become progressively more difficult, with the last few questions being the hardest. On the Chemistry GSE, however, there is no order of difficulty on the multiple-choice section of the exam. So doing the test straight through, from 1 to 30, may not be your best strategy. Your goal on the test is to work as rapidly as you can without sacrificing accuracy. This means that if you find that a question is difficult for you, leave that one for later, and move on to another question.

The Two-Pass System

Have you ever been given a question that stumped you, but you were sure you could answer it? Have you ever said, "Just one more minute—I know I can figure this out!" Well, we all have, and we all know that one more minute sometimes means five more minutes, and often, we don't end up with the right answer at all.

Don't let one question ruin your whole day. You've got a certain number of questions to tackle, and allowing one to throw off your timing might set you back. Here is a general rule for the multiple-choice section: *If you haven't figured out the correct answer in 90 seconds, skip the question and come back to it later.* We're not telling you to give up on it—if you can't answer the question, make a small mark on your answer sheet so you can come back to it later. After you complete the section, go back to the questions you weren't able to solve. Remember to use POE on these questions, and make sure you have selected an answer choice for every question before time is up.

We call this strategy the two-pass system. The first time you go through a section, try every question. If a question seems too difficult or stumps you, move right along. Once you've completed the section, go back to those questions. If you still aren't sure how to solve them, use POE and then make a guess.

Oftentimes, when you go back to a problem a second time you'll have a revelation about how to solve it. (We've all left a test and said, "Oh yeah! Now I know what the answer to number 5 was.") Skipping the problem and then going back to it might give you a chance to have this revelation *during* the test, when it will still be useful.

Follow the Template

Many students think the written response section of the Chemistry GSE is the most difficult part of the exam. Later in the book, we will spend time practicing exactly how to approach these written response questions so that you'll be comfortable with them by the time test day comes. For now, you only need to remember this about answering a written response question: Write out everything you do clearly, and provide explanations!

To receive a high score on the written response section, you will need to provide explanations for the material you present. If you aren't sure about your answer, make sure you explain any rules you used to make your calculations. You can receive a lot of partial credit on this section for providing sound explanations, even if you don't end up with the correct answer. Students who leave things blank because they aren't sure of the answer will cost themselves lots of points on this section. The reverse is also true: Students may give a correct answer but still lose points for not providing a complete explanation. This is one area in which showing your work is not only helpful, it is vital to scoring well.

We will discuss strategies for the written response questions in greater detail as we move through the review of chemistry concepts.

YOU ARE IN CONTROL

We know that taking the Chemistry GSE can be a stressful process. With all this built-up pressure, it might feel like this test is totally out of your control, but the opposite is true—you are in control. Although you can't decide what number pencil to bring to the exam (you must bring a #2) or where to sit during the test, you can decide how you take the GSE. So let's review what we've discussed in this chapter:

- First, you can take advantage of the multiple-choice format of Part I. There is no guessing penalty, and you can use Process of Elimination to add points to your score, even without knowing the correct answer.

- Second, you can answer the multiple-choice questions in any order you want. Spend time with questions that you're comfortable with. If question 12 is really stumping you, move on to question 13 and return to 12 later.

- Third, you can gain points on the written response section by providing clear explanations. Sometimes mentioning a definition or rule can help you gain additional points, even if you do not know the correct answer.

As you build upon your knowledge of chemistry by reviewing the chapters ahead, you'll gain more confidence in your ability to handle the exam.

BAG OF TRICKS SUMMARY

Here is a list of the tricks you'll find in your bag of tricks—be sure to make good use of them.

- An empty Scantron sheet is a bad Scantron sheet
- Process of Elimination (POE)
- You don't have to start at number 1
- The two-pass system
- Follow the template

Now let's begin the chemistry review.

Chapter 3

ATOMIC STRUCTURE AND THE PERIODIC TABLE

An element is a substance that cannot be broken down into any other substance. The periodic table provides you with certain characteristics of the 109 elements and illustrates their chemical properties and trends.

Great news! You do not need to memorize the 109 elements of the periodic table. The periodic table, as depicted below, will appear in your test booklet for your reference.

Periodic Table of the Elements

1 H 1.0																	2 He 4.0
3 Li 6.9	4 Be 9.0											5 B 10.8	6 C 12.0	7 N 14.0	8 O 16.0	9 F 19.0	10 Ne 20.2
11 Na 23.0	12 Mg 24.3											13 Al 27.0	14 Si 28.1	15 P 31.0	16 S 32.1	17 Cl 35.5	18 Ar 39.9
19 K 39.1	20 Ca 40.1	21 Sc 45.0	22 Ti 47.9	23 V 50.9	24 Cr 52.0	25 Mn 54.9	26 Fe 55.8	27 Co 58.9	28 Ni 58.7	29 Cu 63.5	30 Zn 65.4	31 Ga 69.7	32 Ge 72.6	33 As 74.9	34 Se 79.0	35 Br 79.9	36 Kr 83.8
37 Rb 85.5	38 Sr 87.6	39 Y 88.9	40 Zr 91.2	41 Nb 92.9	42 Mo 95.9	43 Te (98)	44 Ru 101.1	45 Rh 102.9	46 Pd 106.4	47 Ag 107.9	48 Cd 112.4	49 In 114.8	50 Sn 118.7	51 Sb 121.8	52 Te 127.6	53 I 126.9	54 Xe 131.3
55 Cs 132.9	56 Ba 137.3	57 *La 138.9	72 Hf 178.5	73 Ta 180.9	74 W 183.9	75 Re 186.2	76 Os 190.2	77 Ir 192.2	78 Pt 195.1	79 Au 197.0	80 Hg 200.6	81 Tl 204.4	82 Pb 207.2	83 Bi 209.0	84 Po (209)	85 At (210)	86 Rn (222)
87 Fr (223)	88 Ra 226.0	89 †Ac 227.0	104 Unq (261)	105 Unp (262)	106 Unh (263)	107 Uns (262)	108 Uno (265)	109 Une (267)									

*Lanthanide Series:	58 Ce 140.1	59 Pr 140.9	60 Nd 144.2	61 Pm (145)	62 Sm 150.4	63 Eu 152.0	64 Gd 157.3	65 Tb 158.9	66 Dy 162.5	67 Ho 164.9	68 Er 167.3	69 Tm 168.9	70 Yb 173.0	71 Lu 175.0
†Actinide Series:	90 Th 232.0	91 Pa (231)	92 U 238.0	93 Np (237)	94 Pu (244)	95 Am (243)	96 Cm (247)	97 Bk (247)	98 Cf (251)	99 Es (252)	100 Fm (257)	101 Md (258)	102 No (259)	103 Lr (260)

The horizontal rows in the periodic table are called **periods.** The vertical columns are called **groups.**

Since the test writers will provide you with this useful tool, it is important you know how to read it. In the next two chapters, we'll review how to read and interpret the periodic table.

THE STRUCTURE OF AN ATOM

The smallest measurable unit of an element is an **atom**, and many atoms bonded together make up a molecule. Just because we can't see an atom or put one on a scale to measure its weight does not mean that an atom has no mass. In fact, an atom's mass plays a key role in determining how it reacts with other atoms in chemical reactions.

What makes up an atom? Well, a **nucleus** sits at the center of every atom. The nucleus contains **protons** and **neutrons** and is the heaviest part of the atom. Each proton carries a charge of +1 while neutrons have no charge. This gives the nucleus an overall positive charge. The mass of a neutron is nearly equal to the mass of a proton.

Outside the positively charged nucleus are negatively charged **electrons**; each electron carries a charge of –1. Electrons are extremely light; their total mass is insignificant in relation to that of the entire atom.

ATOMIC NUMBER, ATOMIC SYMBOL, AND ATOMIC MASS

Let's look at the following example of an element as it appears in the periodic table.

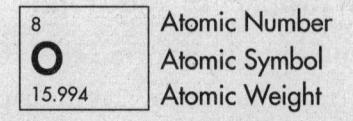

8 O 15.994	Atomic Number Atomic Symbol Atomic Weight

Three important pieces of information are illustrated in the box above.

The letter O represents the **atomic symbol** for this element. O is the atomic symbol that stands for oxygen. Be careful; the atomic symbol is not always the first letter of the element's name. For instance, K stands for potassium and Ag stands for silver!

The number at the top of the box is the **atomic number** of that element. The atomic number is equal to the number of protons in one atom of the element. The atomic number for oxygen is 8, which is the total number of protons in an oxygen atom. If a proton were somehow removed from the nucleus of an oxygen atom to reduce the number of protons to 7, the atom would no longer be oxygen; it would be nitrogen. Therefore, the atomic number *defines* the element.

The atomic number also represents the number of electrons surrounding the nucleus when the atom is in its neutral state. When the number of electrons in an atom equals the number of protons, the sum of the positive and negative charges is zero, and the atom is said to be electrically neutral. When an atom loses or gains electrons, it carries either a negative or positive charge. An atom that carries a net charge is called an **ion**.

Atoms that gain electrons become negatively charged and are called **anions**, while atoms that lose electrons become positively charged and are called **cations**. Atoms gain or lose electrons without changing their identity. In other words, if an oxygen atom were to gain or lose electrons, the resulting ion would still be oxygen.

The number at the bottom of the box represents the **atomic weight** of the element. This is the average weight of an atom that is present in a naturally occurring sample of the element, and is measured in atomic mass units, represented by u. Not all atoms of the same element are created equal. Some atoms will have a different number of neutrons in their nuclei, but if you were to take a large sample of oxygen atoms, the average atomic weight would be 15.994 u.

Another illustration that you should be familiar with is the following:

$$\begin{matrix} \text{Mass Number} \rightarrow 12 \\ \text{Atomic Number} \rightarrow 6 \end{matrix} \text{C}$$

The letter again represents the atomic symbol; C represents carbon. The bottom number is the atomic number, or the total number of protons, and the top number is the **mass number**. The mass number of an element is simply the number of protons plus the number of neutrons. From the information given above, we can determine the number of neutrons that carbon usually contains. If the mass number is 12 and the number of protons is 6, then the total number of neutrons is 6.

Mass number = # of protons + # of neutrons

# of neutrons = mass number – # of protons

# of neutrons = 12 – 6

# of neutrons = 6

Look at the following carbon atoms:

$$^{12}_{6}\text{C} \quad ^{14}_{6}\text{C}$$

Both of these carbon atoms contain the same number of protons, but they have different **mass numbers**, which means that they have a different number of neutrons. They are **isotopes**. Isotopes are denoted by their mass number, followed by their atomic symbol, for instance, ^{12}C and ^{14}C.

ELECTRONS AND QUANTUM THEORY

An atom's **electron configuration** describes how electrons are distributed about the atom. The position of an electron as it orbits an atom's nucleus can also be identified by four **quantum numbers:** n, l, m, and s or m_s. Since it is not possible to tell exactly where an electron is located without disrupting the position or momentum of the electron, the quantum numbers represent only a probability of finding an electron at a certain location.

Since space has three dimensions, the first three quantum numbers are needed to identify a probable spatial location. The first three numbers identify an electron's shell (n), subshell (l), and orbital (m). The fourth quantum number represents the spin (s) of the electron, which reflects its magnetic property. Since no two electrons can be in the same place at the same time, no two will have the same set of four quantum numbers. This is known as the **Pauli exclusion principle**.

First Quantum Number

Shells: n = 1, 2, 3...

Electrons exist in **orbitals** around the nucleus. Orbitals are defined as areas with particular energy and specific quantum number values, and they can be different shapes and sizes. The shell, n, is the principle quantum number and is related to the size and energy of the orbital. It is represented by a positive integer, one or higher. Electrons in larger orbitals are likely to be farther away from the nucleus and have greater potential energy than electrons in smaller orbitals. Therefore, the greater the n value, the larger the shell and the greater the energy of the electron.

Second Quantum Number

Subshells: l = 0, 1, 2..., (n−1)

The second quantum number is the **angular momentum quantum number** (l). It describes orbitals of n that have different shapes. The number can be an integer starting with zero, and it increases by increments of one, up to (n − 1). Orbitals of the same shell, but which have different shapes, are part of different subshells. The four subshells to remember are s, p, d, and f. The first shell (n = 1) contains the s subshell (l = 0). The second shell (n = 2) contains s and p subshells, (l = 0 and l = 1, respectively). The third shell (n = 3) contains the s, p, and d subshells, (l = 0, l = 1, and l = 2). The fourth shell (n = 4) contains subshells s (l = 0), p (l = 1), d (l = 2), and f (l = 3).

Third Quantum Number
Orbitals: m =...−1, 0, +1, ...

The third quantum number is the **magnetic quantum number (*m*)**, which represents the electron's orbital. This value is related to the orientation of the orbital in space and is an integer ranging from −*l* to +*l*, including zero. Exactly how many orbitals are there in a particular subshell? In order to determine the number of orbitals per subshell, multiply the second quantum number (*l*) by two and add one. For example, the *s* subshell (*l* = 0) contains one orbital, *m* = 0. An orbital of an *s* subshell will be spherically shaped.

We can use an orbital diagram to represent the s subshell, and later we can use this diagram to depict the order of addition of electrons to the subshells.

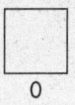

☐
0

The *p* subshell (*l* = 1) contains three *p* dumbbell-shaped orbitals, *m* = −1, *m* = 0, and *m* = +1. The *p* orbitals differ in their alignment in space and are denoted by p_x, p_y and p_z.

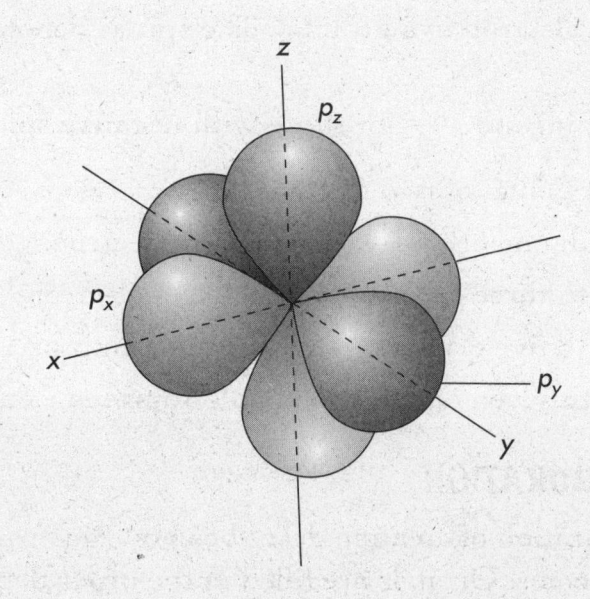

The orbital diagram for the *p* subshell can be drawn like this:

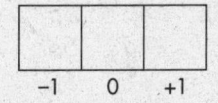

−1	0	+1

ATOMIC STRUCTURE AND THE PERIODIC TABLE ◆ 21

The d subshell ($l = 2$) contains five intricately shaped orbitals, $m = -2$, $m = -1$, $m = 0$, $m = 1$ and $m = 2$.

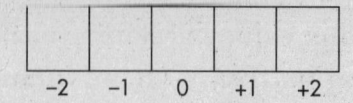

```
| -2 | -1 | 0 | +1 | +2 |
```

The f subshell ($l = 3$) contains seven complexly shaped orbitals, $m = -3$, $m = -2$, $m = -1$, $m = 0$, $m = 1$, $m = 2$ and $m = 3$.

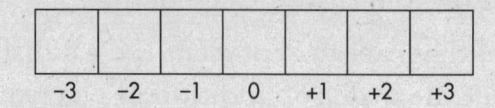

```
| -3 | -2 | -1 | 0 | +1 | +2 | +3 |
```

FOURTH QUANTUM NUMBER

Spin: $s = +\frac{1}{2}, -\frac{1}{2}$

The fourth quantum number designates the **spin** (s) of the electrons in each orbital. Each orbital can contain a maximum of two electrons, and those electrons must have opposite spins. Therefore, there are only two possible spin orientations for each electron in an orbital. The spin is denoted by $+\frac{1}{2}$ for electrons with positive spin, and $-\frac{1}{2}$ for those with negative spin. And since each orbital contains a maximum of two electrons:

the s subshell (with one orbital) will hold two electrons
the p subshell (with three orbitals) will hold six electrons
the d subshell (with five orbitals) will hold ten electrons
the f subshell (with seven orbitals) will hold fourteen electrons

ELECTRON CONFIGURATION

The electron configuration of an atom describes how electrons are distributed around an atom's nucleus. Orbitals are filled in the order depicted in the diagram below.

1s	2s	3s	4s	5s	6s	7s	8s
	2p	3p	4p	5p	6p	7p	
		3d	4d	5d	6d		
			4f	5f			

$1s\ 2s\ 2p\ 3s\ 3p\ 4s\ 3d\ 4p\ 5s\ 4d\ 5p\ 6s\ 4f\ 5d\ 6p\ 7s\ 5f\ 6d$

The electron configuration is denoted by the shell number, subshell symbol, and the number of electrons in the subshell. For example, the electron configuration of a nitrogen atom (atomic number = 7) is $1s^2 2s^2 2p^3$. If we want to denote how electrons fill orbitals, we can use orbital diagrams:

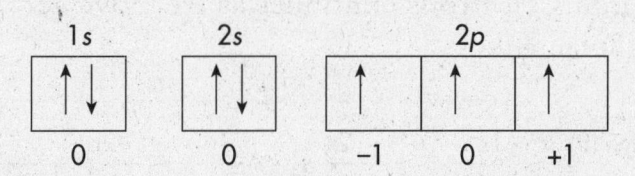

The orbital diagram for a nitrogen atom reveals that there are two electrons in the $1s$ orbital ($m = +\frac{1}{2}$ and $m = -\frac{1}{2}$), two electrons in the $2s$ orbital ($m = +\frac{1}{2}$ and $m = -\frac{1}{2}$), and three electrons in the $2p$ orbital ($m = +\frac{1}{2}$, $m = +\frac{1}{2}$ and $m = +\frac{1}{2}$). Notice how the $2p$ orbital is filled. One positive-spin electron is assigned to each orbital before electrons are paired. This is known as **Hund's rule.** In order to obtain the lowest energy state, electrons with the same spin are added to separate orbitals of the same subshell before pairing together.

In order to denote the quantum numbers for nitrogen, we look at the last electron assigned.

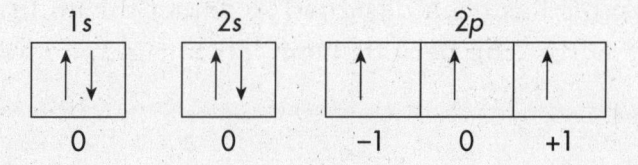

The quantum numbers for this electron would be

$n = 2 \qquad l = 1 \qquad m = +1 \qquad s = +\frac{1}{2}$

The periodic table is arranged so that the sequential addition of electrons to orbitals occurs in the same order as the atomic number. The element to the right of nitrogen is oxygen, which has an electron configuration of $1s^2 2s^2 2p^4$. The orbital diagram for oxygen is:

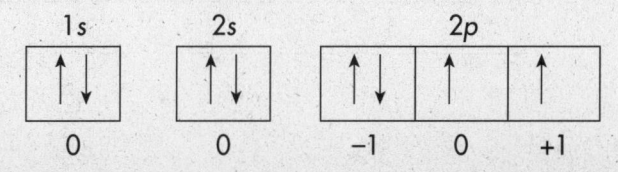

ATOMIC STRUCTURE AND THE PERIODIC TABLE ◆ 23

Since oxygen has one more electron than nitrogen, one electron is added to the orbital diagram to share a p orbital. The four quantum numbers for the oxygen atom are:

$$n = 2 \qquad l = 1 \qquad m = -1 \qquad s = -\frac{1}{2}$$

This pattern of adding electrons continues as we move across the table, completing this period of elements.

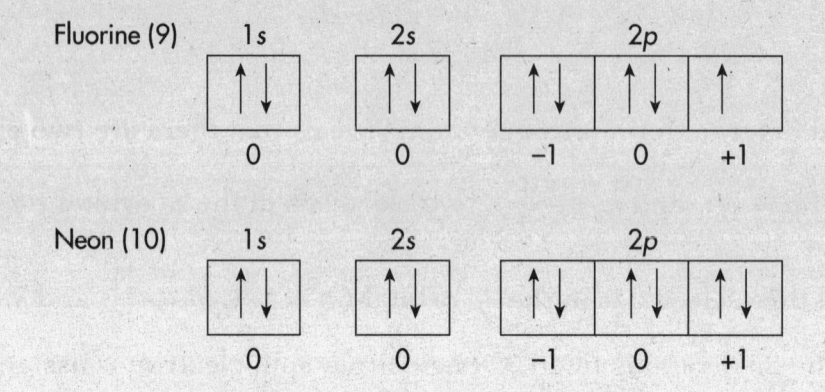

Fluorine (9)
1s	2s	2p
↑↓	↑↓	↑↓ ↑↓ ↑
0	0	−1 0 +1

Neon (10)
1s	2s	2p
↑↓	↑↓	↑↓ ↑↓ ↑↓
0	0	−1 0 +1

Notice that all of neon's orbitals are completely full.

Another way to correctly represent the electron configuration of an atom is by using the **noble gas configuration**; for instance magnesium, Mg, would be Mg = $1s^2 2s^2 2p^6 3s^2$, or [Ne]$3s^2$.

The periodic table has been designed to depict the addition of electrons to orbitals traveling across the periods from left to right, and down each period.

1s				1s
2s				2p
3s				3p
4s	3d			4p
5s	4d			5p
6s	* 5d			6p
7s	** 6d			

*	4f
**	5f

Being able to interpret the periodic table in this way will help you determine the configuration of elements fairly quickly.

Notice that in the above diagram, electrons occupy the *4s* orbital before the *3d* orbital. This occurs again with the *5s*, *6s* and *7s* orbitals.

Watch out! At the point at which the first asterisk appears on the table, after one electron has been added the *5d* subshell, the next electron is added to the *4f* subshell. Electrons continue to add to the *4f* subshell until it is completely filled, before they begin to fill the remaining space of the *5d* subshell. (The elements of the *4f* subshell are known as the **lanthanides**.) Let's look at an example. The electron configuration for terbium (atomic number = 65) is [Xe]$6s^2 5d^1 4f^8$. Notice that this occurs again at the double asterisk (**). This time, after one electron has been added to the *6d* subshell, the next electron is added to the *5f* subshell. The elements of the *5f* subshell are known as the **actinides**.

VALENCE ELECTRONS

The periodic table also helps you determine the configuration of the outermost electrons in an atom; these outermost electrons are also known as **valence electrons**. Valence electrons are involved in chemical reactions. If your periodic table does not already list the group numbers across the top of the table, write them at the top of each column as follows:

1A	2A	3B	4B	5B	6B	7B	8B			1B	2B	3A	4A	5A	6A	7A	8A
1 H 1.0																	2 He 4.0
3 Li 6.9	4 Be 9.0											5 B 10.8	6 C 12.0	7 N 14.0	8 O 16.0	9 F 19.0	10 Ne 20.2
11 Na 23.0	12 Mg 24.3											13 Al 27.0	14 Si 28.1	15 P 31.0	16 S 32.1	17 Cl 35.5	18 Ar 39.9
19 K 39.1	20 Ca 40.1	21 Sc 45.0	22 Ti 47.9	23 V 50.9	24 Cr 52.0	25 Mn 54.9	26 Fe 55.8	27 Co 58.9	28 Ni 58.7	29 Cu 63.5	30 Zn 65.4	31 Ga 69.7	32 Ge 72.6	33 As 74.9	34 Se 79.0	35 Br 79.9	36 Kr 83.8
37 Rb 85.5	38 Sr 87.6	39 Y 88.9	40 Zr 91.2	41 Nb 92.9	42 Mo 95.9	43 Te (98)	44 Ru 101.1	45 Rh 102.9	46 Pd 106.4	47 Ag 107.9	48 Cd 112.4	49 In 114.8	50 Sn 118.7	51 Sb 121.8	52 Te 127.6	53 I 126.9	54 Xe 131.3
55 Cs 132.9	56 Ba 137.3	57 *La 138.9	72 Hf 178.5	73 Ta 180.9	74 W 183.9	75 Re 186.2	76 Os 190.2	77 Ir 192.2	78 Pt 195.1	79 Au 197.0	80 Hg 200.6	81 Tl 204.4	82 Pb 207.2	83 Bi 209.0	84 Po (209)	85 At (210)	86 Rn (222)
87 Fr (223)	88 Ra 226.0	89 †Ac 227.0	104 Unq (261)	105 Unp (262)	106 Unh (263)	107 Uns (262)	108 Uno (265)	109 Une (267)									

*Lanthanide Series:	58 Ce 140.1	59 Pr 140.9	60 Nd 144.2	61 Pm (145)	62 Sm 150.4	63 Eu 152.0	64 Gd 157.3	65 Tb 158.9	66 Dy 162.5	67 Ho 164.9	68 Er 167.3	69 Tm 168.9	70 Yb 173.0	71 Lu 175.0
†Actinide Series:	90 Th 232.0	91 Pa (231)	92 U 238.0	93 Np (237)	94 Pu (244)	95 Am (243)	96 Cm (247)	97 Bk (247)	98 Cr (251)	99 Es (252)	100 Fm (257)	101 Md (258)	102 No (259)	103 Lr (260)

The group number will tell you the number of valence electrons that an atom in that group may have. For example, both oxygen and sulfur are located in group number six, and both oxygen and sulfur have six electrons in their outermost shell.

$$\text{Oxygen} - 1s^2\underline{2s^22p^4}$$

$$\text{Sulfur} - 1s^22s^22p^6\underline{3s^23p^4}$$

The outermost shell that's occupied by electrons in an oxygen atom is $n = 2$. The $2s$ subshell has two electrons and the $2p$ subshell has four, so oxygen has a total of six valence electrons. The outermost shell that's occupied by electrons in a sulfur atom is $n = 3$. The $3s$ subshell contains two electrons and the $3p$ subshell contains four electrons for a total of six valence electrons. Later we will review how valence electrons play a role in atomic bonding.

MAGNETIC PROPERTIES

Let's take a look at the orbital diagram for neon again.

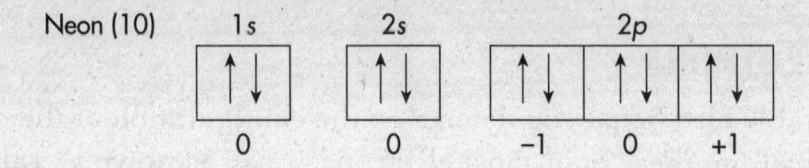

```
Neon (10)    1s        2s            2p
            ┌──┐      ┌──┐      ┌──┬──┬──┐
            │↑↓│      │↑↓│      │↑↓│↑↓│↑↓│
            └──┘      └──┘      └──┴──┴──┘
             0         0        -1   0  +1
```

As we said before, all of neon's orbitals are filled, and all of its electrons are spin paired. An element that has all of its electrons spin paired is a **diamagnetic element**. Other examples of diamagnetic elements are:

Helium	$1s^2$
Magnesium	$1s^2\ 2s^2\ 2p^6\ 3s^2$
Argon	$1s^2\ 2s^2\ 2p^6\ 3s^2\ 3p^6$

An atom of a diamagnetic element is not attracted by a magnetic field. In fact, atoms of diamagnetic elements may be slightly repelled by magnetic fields.

Most elements do not have orbitals that are completely full, and they are called **paramagnetic elements**. Paramagnetic elements are attracted by magnetic fields.

OTHER THEORIES

The Heisenberg Uncertainty Principle

Since electron orbitals only represent the *probability* of finding an electron at a specific point in space at any moment, it is impossible to know both the position and momentum of an electron at any time. This concept is known as the **Heisenberg uncertainty principle**.

DeBroglie Hypothesis

Louis deBroglie postulated that light has wave characteristics. You can think of a light wave as a stream of particles oscillating through space. Light energy travels in waves of electromagnetic radiation. Waves are measured by both their wavelength (λ) and frequency (*f*). **Wavelength** (λ) is the length between two adjacent wave peaks (or troughs) and represents the distance of one full cycle of a wave. Wavelength is measured in nanometers (nm).

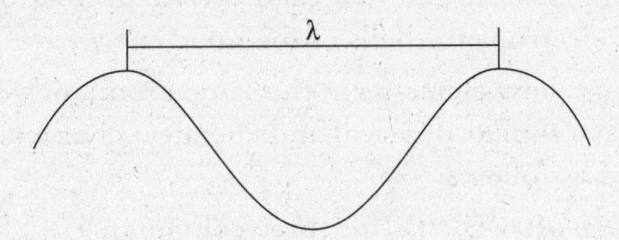

Frequency (*f*) is the number of wave cycles that pass by a certain point in 1 second. The unit of frequency is called the hertz (Hz).

The product of wavelength and frequency equals the number of complete waves that have passed a certain point in 1 second. This is known as the wave's **velocity**. Light waves differ in frequency and wavelength but always have a velocity of 3×10^8 m/s. Therefore,

$\lambda f = c$

where *c* is the speed of light.

According to the equation above, wavelength and frequency are inversely proportional. This means that as wavelength *increases*, frequency *decreases*.

The rate at which light travels depends on the medium through which it travels. The speed of light is equal to 3×10^8 m/s only if light travels through a vacuum. Traveling through air, the speed of light is a little less than this value, but for any calculations you see on the GSE, you should assume that light is traveling through a vacuum.

Electrons also exhibit wave characteristics. The **de Broglie hypothesis** states that there is a relationship between an electron's mass, speed, and wavelength. This relationship is represented in the formula below.

$\lambda = h/mv$

Where λ = associated wavelength

h = Planck's constant (6.63×10^{-34} joule-second)

m = mass of the particle

v = speed of the particle

As we mentioned, the de Broglie hypothesis applies to calculations involving electrons. Electrons all have the same mass but will differ in velocity and wavelength. According to the De Broglie formula, velocity and wavelength are also inversely proportional. That is, as velocity increases, wavelength decreases.

REVIEW OF THE PERIODIC TABLE

We discussed earlier how elements on the periodic table are divided into four areas: *s*, *p*, *d*, and *f*. We also learned that each period, or horizontal row, represents the addition of electrons to shells of the same energy.

We will now discuss how elements in the same group, or vertical column on the periodic table, have similar physical and chemical characteristics. The groups are numbered as follows:

1A, 2A, 3B, 4B, 5B, 6B, 7B, 8B (for three columns), 1B, 2B, 3A, 4A, 5A, 6A, 7A, 8A

$$\text{Mass Number} \rightarrow 12 \atop \text{Atomic Number} \rightarrow 6 \text{ C}$$

Elements within the same group have the same number of electrons in their outermost shell. For instance, elements of the first group (IA) on the periodic table all have one valence electron placed in an *s* subshell. With the exception of hydrogen, atoms of elements in this group have similar characteristics, and all are very reactive. All of these elements are also metals that have similar appearances. This group of elements, from lithium (Li) to francium (Fr) is called the **alkali metals**.

Elements of the second group (IIA) all have two valence electrons. Atoms of these elements are reactive, but not as reactive as those in the first group. Members of this group, from beryllium (Be) to radium (Ra) also have similar appearances and are called the **alkaline earth metals**. The first two groups are also referred to as the **active metals**, since both groups are much more reactive than other common metals of the periodic table.

Now let's look at a couple of other important groups of elements, starting with VIIIA. All of these elements, excluding helium (He), have eight valence electrons and are diamagnetic. Atoms from this group of elements known as the **noble gases** are generally unreactive. To the left of the noble gases, in group VIIA, are the **halogens**. Atoms from this group of elements all have seven valence electrons and are extremely reactive.

METALS, NONMETALS, AND SEMI-METALS

Most of the elements on the periodic table are **metals**. Metals make up approximately 75 percent of all elements and have many similar physical properties in common. They have a characteristic shiny appearance and easily conduct heat and electricity. Most metals are also malleable, which means they are capable of being shaped by pressure, and ductile, which means that they can be drawn into wire. Metals also have chemical properties in common, and they generally lose electrons when they are engaged in bonds. But some metals are more reactive than others. As we discussed earlier, the **active metals,** which comprise the first two groups of elements, are relatively very reactive, while the other metals on the periodic table, located in the ∂ and f regions, are referred to as the **transition metals**. These metals are less reactive and, for the most part, are harder than the active metals.

Elements that are classified as nonmetals are found on the right-hand side of the periodic table. Although most of the nonmetals are gases at room temperature (such as oxygen and fluorine), some are solids (like phosphorous and sulfur). Only one nonmetal, bromine, is a liquid at room temperature. Not surprisingly, nonmetals do not have the same properties as metals. They do not have a lustrous, shiny appearance, and they don't conduct heat or electricity very well. When involved in chemical bonds, nonmetals generally share or even gain electrons.

Lastly, seven elements on the periodic table are classified as **metalloids**, or **semi-metals**. These elements have both metallic and nonmetallic properties. The periodic table below reveals the classifications of the elements as metals, metalloids, or nonmetals.

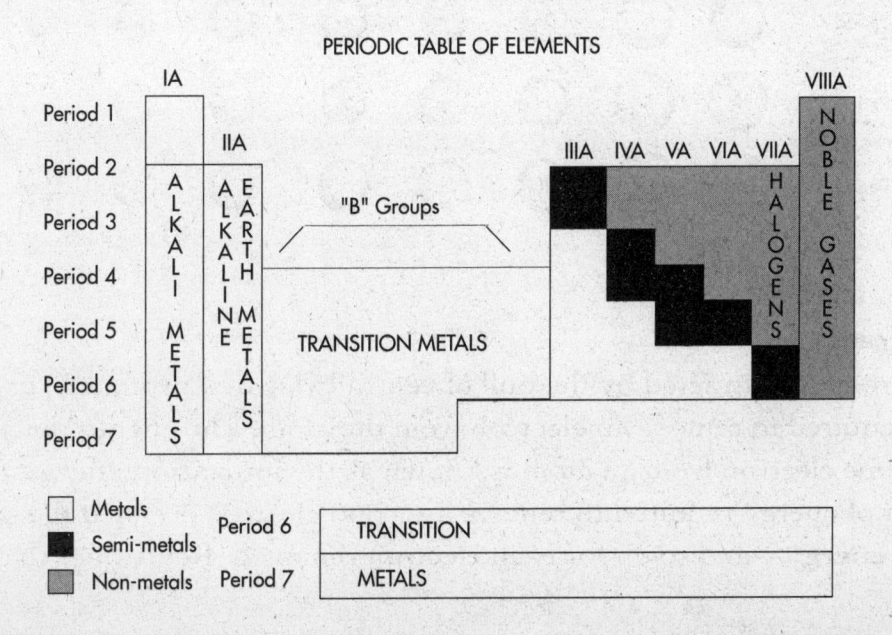

PERIODIC TABLE OF ELEMENTS

PERIODIC TRENDS

Atomic Radius

The positive charges of protons in the nucleus of an atom exert a pull on the negatively charged electrons that surround the nucleus. However, the shells that exist between the nucleus and the valence electrons act as shields against this pull in an effect known as **nuclear shielding.** The amount or degree of nuclear shielding is directly related to the size of an atom. With each new shell that's added to an atom, the valence electrons are further shielded from the attractive force of the protons. If additional protons are added without the addition of a shell (i.e., if they are added to the same shell), the pull of the protons on the electrons *increases* in strength. Therefore, as you go across the periodic table from left to right, the number of protons increases without the addition of a new shell, and since the valence electrons are pulled in more tightly, the size of the atom *decreases*. However, as you go *down* the periodic table, new shells are added, causing the nuclear shielding effect. The size of an atom, therefore, increases as you proceed *down* the periodic table.

	IA	IIA	IIIA	IVA	VA	VIA	VIIA	VIIIA
Period 1	H							He
Period 2	Li	Be	B	C	N	O	F	Ne
Period 3	Na	Mg	Al	Si	P	S	Cl	Ar
Period 4	K	Ca	Ga	Ge	As	Se	Br	Kr
Period 5	Rb	Sr	In	Sn	Sb	Te	I	Xe
Period 6	Cs	Ba	Tl	Pb	Bi	Po	At	Rn

Ionization Energy

Since electrons are attracted by the pull of centrally located protons in an atom, energy is required to remove an electron from the atom. The energy that's needed to remove one electron from an atom is known as the **ionization energy (IE)**, and the amount of energy required to remove a second electron is called the **second ionization energy**—and so on for each electron removed. Remember that when an electron is removed from an atom, the atom's net charge becomes positive; it becomes a cation. The second ionization energy is always greater than first, since the second electron is being taken from a cation, which does not want to lose further valence electrons. For each electron that's removed, the attraction on the remaining valence electrons becomes greater: The ionization energy increases with each subsequent removal. Once a shell has been completely emptied, the ionization energy for the removal of the next electron increases significantly since it must be removed from a shell that's closer to the nucleus.

Ionization energy increases as you go across the periodic table from left to right, since valence electrons are becoming more tightly bounded to the atom as we mentioned in the last section. But ionization energy decreases as we go *down* a group, because the shielding effect causes valence electrons to be less tightly held.

Electron Affinity

Electron affinity refers to the amount of energy that's either released or absorbed when an electron is added to a neutral atom. In the cases of the noble gases and the alkaline earth metals, which have completely filled outer subshells, adding an electron will not create stable negative ions. For elements in groups IIA and VIIIA, energy is required to add an electron, since the electron must be added to a higher energy level. For the majority of the elements, adding an electron makes an atom more stable, since their outer orbitals are not full. For these atoms, energy is released during the addition of an electron.

As you go across the periodic table from left to right, electron affinities generally increase. As you go down a group, electron affinities generally decrease, but not by a significant amount.

Electronegativity

The measure of the nucleus' pull on the electrons of another atom when involved in a chemical bond is called electronegativity. Just as is the case with ionization energy, the electronegativity of an atom increases as you go across the periodic table, and decreases as you go down a group.

Metallic Character

An atom's readiness in giving up an electron to become a positive ion (a cation) is a measure of its metallic character. A highly metallic character means a high degree of valence mobility. (This is why metals are such good conductors!) As we move across the periodic table from left to right, valence electron mobility decreases (as does metallic character) since the valence electrons are more tightly held.

E Rule

The E rule can help you remember all of the trends of the periodic table. All of the terms that start with the letter E increase in the direction of the arrow.

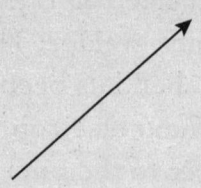

As you move *up* a column and across a row (from left to right), the following properties increase:

- Electronegativity
- Electron affinity
- Ionization energy

The remaining properties follow the reverse order and decrease as you move up a column and across a row. These properties are:

- Atomic radius
- Metallic character

Chapter 4

BONDING

An atom's valence electrons play a significant role in determining the types of bonds it forms. A bond is an equal or unequal sharing of electrons between two atoms. Atoms come together to form molecules in order to gain stability; optimal stability is obtained when an atom's outermost shell contains eight electrons, known as the **octet rule**. The electron configuration of this stable octet is ns^2np^6, also known as a **noble gas configuration**. Unstable atoms have either fewer or more electrons in their outermost shells and form bonds to achieve this stable octet. (Obviously, some elements cannot achieve this type of stability. For example, hydrogen can have a maximum of two electrons in its outermost shell.)

LEWIS DOT STRUCTURES

On the Chemistry GSE, it will be very important for you to understand how to draw Lewis dot structures. A **Lewis dot structure** is merely a representation of an atom's valence electrons. Consider the following chlorine atom:

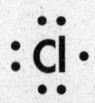

If you look at the periodic table, you can see that chlorine has seven electrons in its outermost shells. These valence electrons are the bonding electrons, and are depicted by dots in the Lewis dot structure above. Six of these bonding electrons are paired, while the seventh is alone.

COVALENT BONDS

Covalent bonds are a result of the attraction between two atomic nuclei that results from their sharing a pair of electrons. Let's take a look at two chlorine atoms:

$$:\ddot{\text{Cl}}\cdot \quad \cdot\ddot{\text{Cl}}:$$

Since a lone chlorine atom has an unpaired electron, it's not very stable. This means that it wants to bond with another atom in order to complete its outermost valence shell. When two chlorine atoms come together, each will share its unpaired electron with the other atom, and the resulting bond is called a covalent bond.

Atoms often form as many covalent bonds as they have unpaired electrons in their outermost shell. For example, let's look at a nitrogen atom:

$$\cdot \dot{\text{N}} :$$

According to its position on the periodic table, nitrogen has five valence electrons, three of which are unpaired. With its three unpaired electrons, nitrogen can form three covalent bonds, as it does in the NH_3 (ammonia) molecule.

$$\begin{matrix} & H & \\ & \cdot\cdot & \\ H & \cdot\cdot N : & \\ & \cdot\cdot & \\ & H & \end{matrix} \quad \text{becomes} \quad \begin{matrix} & H & \\ & \cdot\cdot & \\ H & : N : & \\ & \cdot\cdot & \\ & H & \end{matrix}$$

Each of the three hydrogen atoms can share its unpaired electron with one of nitrogen's unpaired electrons.

MULTIPLE BONDS

The covalent bonds between nitrogen and hydrogen in the ammonia molecule are single bonds, meaning that two electrons are shared between two atoms. But sometimes two atoms share more than one pair of electrons. For instance, a **double bond** occurs when two atoms share *two* pairs of electrons, and a **triple bond** occurs when two atoms share *three* pairs of electrons.

The carbon dioxide molecule (CO_2) is an example of a molecule containing double bonds.

$$\ddot{O}=C=\ddot{O}$$

The hydrogen cyanide molecule (HCN) is an example of a molecule that contains one triple bond:

$$H-C\equiv N:$$

COORDINATE COVALENT BONDS

What happens if a covalent bond is formed and the electrons shared in the bond were originally provided by just one of the atoms? Well, let's look at the following ammonium ion (NH_4^+):

The hydrogen ion (H^+) is bonded to the nitrogen atom through a covalent bond. Since both of the paired electrons that are now being shared between nitrogen and the fourth hydrogen were originally donated by nitrogen, this bond is referred to as a **coordinate covalent bond**.

POLAR COVALENT BONDS

Remember when we told you that electronegativity is the measure of an atom's pull on electrons when it's involved in a covalent bond? Well, in covalent bonds between two atoms of very different electronegativity, the shared electrons will spend more time rotating around the nucleus of the more electronegative atom. Let's look at a molecule of hydrogen chloride (HCl):

$$H-\ddot{\underset{..}{Cl}}:$$

We can see from its location on the periodic table that chlorine is more electronegative than hydrogen. When these atoms covalently bond, chlorine exerts a stronger pull on the electrons than does hydrogen. As a result, the chlorine side of the molecule becomes more electronegative and the hydrogen side becomes more electropositive. This separation of charge is known as **polarity** and can be represented as follows:

$$\overset{\delta^+\ \ \delta^-}{H-Cl}$$

The bond between the chlorine and hydrogen atom is called a **polar covalent bond**.

The amount of charge separation between two atoms is measured by the **dipole moment** of the molecule. Every polar molecule has a dipole moment that's equal to the value of the partial charge (δ) multiplied by the distance between the charges (∂).

$$\mu = \delta\partial$$

The unit of the dipole moment is coulombs-meters and, not surprisingly, as the polarity of a molecule increases, so does the dipole moment.

A molecule's polarity is also related to its geometric shape. Let's look at that molecule of carbon dioxide again:

$$\overset{\delta^-\ \ \ \delta^+\delta^+\ \ \ \delta^-}{O=C=O}$$
$$\underset{\longleftarrow\ \ \longrightarrow}{}$$

(Notice that partial charges are depicted by δ^+ and δ^-.) The arrows underneath the CO_2 molecule indicate the *direction* of the dipole moment, starting from the partial positive side and heading towards the partial negative side. Since the two bonds are of equal polarity and the dipole moments are in the opposite directions, the **net dipole moment** of this CO_2 molecule is zero.

Another example of the way in which a molecule's geometric shape plays a role in determining its overall dipole moment is seen with water (H₂O).

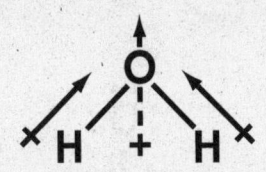

The dipole moment is headed toward the partially negative oxygen atom from the two partially positive hydrogen atoms. Since the dipoles do not negate each other by pointing in opposite directions, the net direction of the dipole can be depicted by the broken arrow above.

WRITING LEWIS DOT STRUCTURES

Here are the steps involved in drawing the Lewis dot structure for the NO_3^- ion.

STEP 1: Determine the total number of valence electrons for the entire molecule or polyatomic ion. Look at each atom's position on the periodic table to determine its number of valence electrons, and remember that *group number equals number of valence electrons*. Nitrogen has five valence electrons and oxygen has six. Since NO_3^- is a polyatomic *anion*, we also need to add one additional electron to the equation. For polyatomic cations like NH_4^+, you need to subtract one electron from the total.

So the number of valence electrons for NO_3^- is 5 + 6 + 6 + 6 + 1 = 24.

STEP 2: Draw the basic structure of the molecule or polyatomic ion—just as the atoms would be situated in space, but adding no electrons or bonds. In our example, nitrogen is the central atom, and it is bonded to three oxygen atoms.

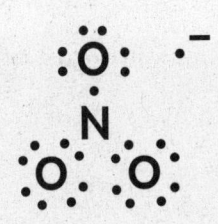

BONDING ◆ 37

STEP 3: Add two electrons to each bond (designated by a line), representing a shared electron pair. If we add all 24 electrons to our diagram, we end up with the following:

(The central nitrogen atom, however, does not have a stable octet and we are fresh out of electrons!)

STEP 4: Now make sure that all of the atoms attached to the central atom have completed octets. If necessary, add electrons in pairs. Then place any remaining electrons on the central atom in pairs. If the central atom doesn't have an octet, then form double or triple bonds.

So, in order for our central nitrogen atom to have eight electrons, it needs to share one more pair of electrons with one of the oxygen atoms in a double bond.

Since this double bond can exist with any of the oxygen atoms, the NO_3^- ion has three **resonance forms**.

The three resonance forms of the NO_3^- ion exist simultaneously in solution.

EXCEPTIONS TO THE RULES

We've been talking about an atom's tendency to form bonds in order to obtain a stable octet but, as we mentioned, this is not possible for elements like hydrogen and beryllium. Both of these atoms cannot obtain eight electrons in their valence shells since they only have *s* subshells. Hydrogen's configuration is $1s^1$ and beryllium's is $1s^2 2s^1$. Since the maximum number of electrons that an *s* subshell can hold is two, both the hydrogen and beryllium atoms will be stable with just two electrons. Boron is another exception to the stable octet rule. Boron bonded to three other atoms can be stable; this gives it six valence electrons. Take a look at the BF_3^- molecule, for example:

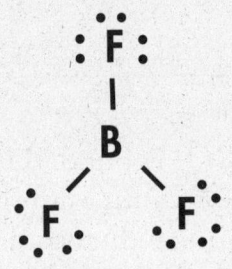

MOLECULAR GEOMETRY

As we saw earlier, carbon dioxide is a linear molecule with a net dipole moment of zero. Water, on the other hand, has a net dipole moment and is bent, not linear. But why is a water molecule bent while carbon dioxide is straight? Well, the number of valence electrons an atom possess also determines its **molecular geometry**. Since all electrons possess a negative charge, they want to be as far away from each other as possible—negative charges repel each other. So electrons arrange themselves in space as far apart as they can, and this dictates a molecule's geometric shape. In order to predict the shapes of molecules, we can use the **valence-shell electron pair repulsion (VESPR) model**.

VESPR MODEL

Okay, so we know that the number of electron domains (A double bond contains two electron pairs, but only counts as one electron "domain.") around a central atom will determine its overall geometry, but other factors have an effect on a molecule's geometry, including the presence of lone pairs of electrons and double and triple bonds. Since lone pairs of electrons have a slightly more repulsive nature than electrons that are bonded, lone pairs take up more space around a central atom. In the same way, double and triple bonds have slightly more repulsive natures than do single bonds, and will also take up more space.

If there are two electron domains on the central atom, then there is only one possible shape for the molecule.

Number of Electron Domains on the central atom	Number of Lone Electron Pairs around the central atom	Example	Molecular Shape	Geometric Diagram
2	0	CO_2	Linear	O = C = O

The central carbon atom in CO_2 is bonded to two atoms and has no unshared pair of electrons. The two bonded atoms want to be situated at opposite sides of the central atom in order to minimize the repulsive forces between them. This results in a linear shape, in which the bond angles are equal to 180°. This molecule also has *sp* hybridization.

If there are three electron domains around the central atom, then two possible shapes arise for the molecule.

Number of Electron Pairs on the central atom	Number of Lone Electron Pairs around the central atom	Example	Molecular Shape	Geometric Diagram
3	0	BF_3	Trigonal Planar	(BF₃ structure)
3	1	SO_2	Bent	(SO₂ structure)

The central boron atom in a BF_3 molecule is bonded to three atoms, and there are no unshared pairs of electrons around it. It is trigonal planar in shape, with bond angles of 120°. The sulfur atom in an SO_2 molecule is bonded to two oxygen atoms and this molecule has one unshared pair of electrons. The SO_2 molecule is bent, with bond angles of approximately 116°. Both of these molecules are sp^2 hybridized.

If there are four electron domains around the central atom, then the molecule can generally assume three shapes.

Number of Electron Domains on the central atom	Number of Lone Electron Pairs around the central atom	Example	Molecular Shape	Geometric Diagram
4	0	CH_4	Tetrahedral	H–C(H)(H)(H)
4	1	NH_3	Trigonal Pyramidal	N(H)(H)(H)
4	2	H_2O	Bent	O(H)(H)

The central carbon atom in methane (CH_4) is bonded to four hydrogen atoms. The molecule is tetrahedral, with bond angles of 109.5°. When looking at the orbital diagram for a carbon atom, you might think that carbon would bond to only two hydrogen atoms.

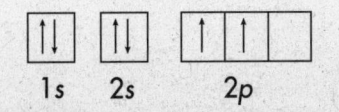

1s ↑↓ 2s ↑↓ 2p ↑ ↑ _

But CH_2 is very reactive, and methane is a molecule that exists in nature. Carbon generally forms four bonds because during bond formation, the electron from its $2s$ is excited to the open $2p$ orbital.

This molecule, therefore, has an sp^3 hybridization.

The nitrogen atom in NH_3 is bonded to three hydrogen atoms and has one lone pair of electrons. The molecule is trigonal pyramidal, with bond angles of approximately 107°. The oxygen in a water molecule (H_2O) is bonded to two hydrogen atoms and has two lone pairs of electrons. Its shape is bent, with bond angles of approximately 105°. Both NH_3 and H_2O are sp^3 hybridized.

Central atoms can also have five and six electron pairs. The following chart depicts all the possible geometries of these types of molecules.

Number of Electron Domains on the central atom	Number of Lone Electron Pairs around the central atom	Example	Molecular Shape	Geometric Diagram
5	0	PCl_5	Trigonal Bipyramidal	
5	1	SF_4	Seesaw or Distorted Tetrahedron	
5	2	ClF_3	T-Shaped	
5	3	XeF_2	Linear	
6	0	SF_6	Octahedral	
6	1	IF_5	Square Pyramidal	
6	2	XeF_4	Square Planar	

IONIC BONDS

Atoms can also bond by completely giving up their electrons to other atoms or by receiving another atom's electrons. Atoms with significantly different electronegativities form these types of bonds, which are called **ionic bonds**. Ionic bonds are formed when the electronegativity difference between two atoms is greater than 1.7. The more electronegative atom steals electrons from its bonding partner.

In ionic bonds, atoms become ions. The atom with the greater electronegativity has acquired additional electrons and becomes an anion, while the atom that has lost its electrons becomes a cation. So what holds the two ions together? The electrostatic attraction between the positively and negatively charged ions causes them to stick together. Finally, a large number of ions together will form an **ionic solid**. Ionic solids generally have a crystalline structure.

Let's look at the bond that forms between sodium and chlorine. As we can tell from their location on the periodic table, there's a big difference in the electronegativity between sodium and chlorine. Chlorine's electron configuration is $[Ne]3s^23p^5$. This means that it has seven valence electrons and is looking to acquire one more electron to complete its valence shell. Sodium's electron configuration is $[Ne]3s^1$. It has only one valence electron, so it's looking to get rid of it and gain a stable noble gas configuration. Chlorine, being the more electronegative atom, will take the electron away from sodium in order to complete its shell. Two ions are formed from this exchange, and these ions (which have opposite charges) are strongly attracted to each other.

ATTRACTIONS BETWEEN MOLECULES

Molecular Solids and van der Waals Forces

Molecular solids are comprised of atoms or molecules that adhere together because of intermolecular forces. There are two types of intermolecular forces, dipole-dipole forces and London dispersion forces (a.k.a. van der Waals forces).

Dipole-dipole forces are attractive intermolecular forces between polar molecules. These forces occur between the electropositive end of one polar molecule and the electronegative end of another. Dipole-dipole forces are fairly weak, and substances held together by dipole-dipole attractions generally have lower melting and boiling points.

London dispersion forces are extremely weak attractive forces between nonpolar molecules. But how can a neutral molecule attract another neutral molecule? Well, the electrons that are orbiting the nucleus of a molecule can move sporadically at times. This means that when a majority of the electrons are at one end of the molecule, a slight polarity may be temporarily induced in a neutral molecule. The frequency of London dispersion forces is greater in molecules that have more electrons, and substances in which these weak London dispersion forces occur generally have lower melting and boiling points.

Substances that contain dipole-dipole forces are typically liquids or gases at room temperature, while substances held together by London dispersion forces are usually gases at room temperature. Dry ice (which is actually carbon dioxide—CO_2) and liquid oxygen are examples of substances that contain van der Waals forces.

Ionic Solids

Ionic solids are comprised of cations and anions held together by the electrostatic attractions between them. Ionic solids have a lattice structure. Consider the molecule of NaCl we looked at earlier. In the crystalline structure of solid NaCl, the positively charged sodium ion (Na^+) attracts six negatively charged chlorine ions (Cl^-) that gather around it. In the same way, every negatively charged chlorine ion attracts six sodium ions that gather around it. The strong bonds between these ionic molecules that form the solid structure and are identical to the strong bonds that hold together the individual ions. Ionic solids (such as salts) generally melt and boil at high temperatures.

Covalent Network Solid

Covalent network solids are comprised of atoms that are held together in a network of covalent bonds. The covalent bonds between the network's molecules are identical to the covalent bonds that hold together the individual atoms. Network solids are extremely strong and have high melting and boiling points. An example of a covalent network solid is diamond. In diamond, a carbon atom is covalently bonded to four other carbon atoms. Diamond is what's known as a three-dimensional covalent network solid, while graphite is an example of a two-dimensional network solid. Graphite is comprised of "sheets" of carbon atoms held together by covalent bonds, but the sheets themselves are held together by van der Waals forces. This means that graphite breaks relatively easily in the horizontal plane, but not in the vertical plane. Asbestos is an example of a one-dimensional covalent network solid; asbestos is comprised of carbon chains held together by covalent bonds.

Metallic Solids

Metallic solids are comprised of positively charged nuclei floating in a sea of moving electrons. These solids have a luster commonly associated with metals. Some examples of metallic solids include copper, iron, gold, and silver. Because of the electron mobility within the solid, metallic bonds are very delocalized. This allows them to be malleable (able to be hammered into sheets), ductile (able to be drawn into wire), and good conductors of heat and electricity. The strength of metallic solids varies, but they are generally fairly hard. Mercury is the only metallic substance that is not a solid at room temperature—it's a liquid.

Hydrogen Bonding

Hydrogen bonds are attractive forces that occur between the partially positive hydrogen atoms of one covalently bonded polar molecule, and the partially negative ends of a covalently bonded molecule bearing an electronegative atom (*N*, *O*, or *F*). Hydrogen bonds are stronger than dipole-dipole forces. When a hydrogen atom loses its electron in a bond, its positive nucleus is essentially unshielded. An example of a substance held together by a bunch of hydrogen bonds is water. Hydrogen bonds are usually represented by a series of dots or dashes.

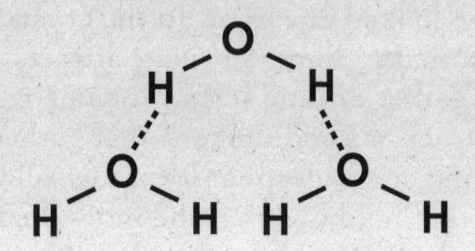

Chapter 5

STOICHIOMETRY

By this time, you are aware that atoms bond together to form molecules. Another term you should be familiar with for the chemistry GSE exam is **molecular formula**. A molecular formula specifies the quantities and identities of the atoms that make up a molecule or compound. For example, the molecular formula for water is H_2O. This tells us that a water molecule is comprised of two hydrogen atoms and one oxygen.

From the molecular formula, we can determine a molecule's formula weight, otherwise known as its **molecular weight (MW)**. Molecular weight is the sum of all the atomic weights of the atoms that make up a molecule. For example, the molecular weight of water is:

$$MW\ H_2O = 2(H) + 1(O)$$

$$MW\ H_2O = 2(1) + 1(16)$$

$$MW\ H_2O = 18$$

The **empirical formula** is the most reduced molecular formula of a compound; it's the molecular formula written with the smallest integer subscripts. Reducing molecular formulas to empirical formulas is just like reducing fractions. When you reduce the fraction $\frac{2}{6}$ to its smallest form, you divide each number by the largest integer that both numbers go into both evenly. In this case, we divide both numbers by two and end up with a reduced fraction of $\frac{1}{3}$. This method for reducing can be used for molecular formulas.

Let's try reducing the formula for ethane (C_2H_6). If you divide both numbers by the largest number that goes into both of them (2), you get the empirical formula CH_3. If a molecule's formula cannot be further simplified, like water (H_2O), then its molecular formula is also its empirical formula.

IN CHEMISTRY, A MOLE IS NOT A SMALL, BURROWING MAMMAL

Even a teaspoon-sized sample of any chemical substance is composed of an immense quantity of molecules or ions. So whenever you're talking about a *visible* quantity of atoms, you're necessarily talking about numbers of atoms that are too large to be counted. For this reason, chemists use the concept of the **mole** to greatly simplify their calculations.

Let's start with a few examples to illustrate the concept of the mole. How many carbon atoms are there in a 12 gram sample of pure carbon-12? There are 6.022×10^{23} carbon atoms. How many Fe atoms are there in a 55.85 gram sample of iron? 6.022×10^{23}. You see, the molecular weight (listed on the periodic table, remember) represents the weight of a sample of the element that contains 6.022×10^{23} atoms of that substance. This number is known as Avogadro's number. A mole of any substance is comprised of Avogadro's number of molecules of that substance. You'll find molar calculations easier if you round **Avogadro's number** to 6.0×10^{23}.

One very important thing for you to remember when you look at compounds is that you have to translate molar ratios according to molecular formula. For instance, one mole of H_2 is comprised of 6.0×10^{23} H_2 molecules, but in this sample there are $2 \times 6.0 \times 10^{23}$ individual H atoms. Think about it this way—there are 6.0×10^{23} molecules of H_2, but it takes two individual H atoms to make up each molecule of H_2, so the number of actual H molecules is 1.20×10^{24}.

Avogadro's number is a very important number in chemistry calculations. Written mathematically, the relationship between moles, molecules, and Avogadro's number is as follows:

$$\text{Moles} = \frac{\text{Molecules}}{6.0 \times 10^{23}}$$

As we mentioned, the atomic weights listed on the periodic table are provided in atomic mass units, and the atomic weights listed also tell you the number of grams per mole of each element. Molecular weight, therefore, is also equal the number of grams per mole of substance. For example, if a water molecule weighs 18 amu, then 1 mole of water molecules weighs roughly 18 grams. The following equation represents the relationship between moles, grams, and molecular weight:

$$\text{Moles} = \frac{\text{Grams}}{\text{Molecular Weight}}$$

MASS COMPOSITION

Using a molecule's empirical formula, we can calculate the percent composition by mass of each element within the molecule. Let's look at a molecule of methane (CH_4). The total molecular weight of methane is:

$$\text{MW } CH_4 = 1(C) + 4(H)$$

$$\text{MW } CH_4 = 12 + 4(1)$$

$$\text{MW } CH_4 = 16$$

The percent mass composition of carbon in methane = $\frac{12}{16}$ = 75%

The percent mass composition of hydrogen in methane = $\frac{4}{16}$ = 25%

If we only know the percent mass composition of the elements in a compound, we can use this to work backwards to figure out a molecule's empirical formula. For example, say a molecule contains 9 percent magnesium (Mg) and 91 percent iodine (I). How do we figure out the molecule's empirical formula from this information? Let's assume that we have a 100 gram sample. (When dealing with percentages, 100 is the easiest number to work with.)

Now, in a 100 gram sample, we would have:

$$9\% \text{ Magnesium} = 9 \text{ g of Mg } (.09 \times 100 = 9)$$

$$91\% \text{ Iodine} = 91 \text{ g of I } (.91 \times 100 = 91)$$

Next convert grams to moles:
(Molecular weight of Mg = 24.31 g/mol)
(Molecular weight of I = 126.9 g/mol)

$$9\text{g Mg}\left(\frac{1 \text{ mol}}{24.31\text{g}}\right) = .37 \text{ moles Mg}$$

$$91g\ I\left(\frac{1\ mol}{126.9g}\right) = .72\ moles\ I$$

Next figure out the molar ratio of magnesium to iodine in the compound. The ratio is 0.72 mol I to 0.37 mol Mg, which is about equal to 2:1. This means that for every two atoms of I in the compound, there should be 1 atom of iodine, which makes the empirical formula MgI_2.

CHEMICAL EQUATIONS

Take a look at the following chemical equation:

$$1N_2 + 3H_2 \rightarrow 2NH_3$$

This equation represents the reaction of diatomic hydrogen (H_2) and nitrogen (N_2) to produce ammonia (NH_3). Both hydrogen (H_2) and nitrogen (N_2) are located on the left side of the equation and are known as the **reactants**. Ammonia (NH_3) is located on the right side of the equation and is called the **product**. The coefficients, which are bolded (1, 2, and 3), are located to the left of each compound and represent the total number of molecules that participate in the reaction. They also represent the molar ratio of the compounds involved in the reaction. So, in the equation above, 1 mole of nitrogen (N_2) reacts with 3 moles of hydrogen molecules (H_2) to produce 2 moles of ammonia (NH_3).

The coefficients represent the molar ratios between products and reactants, and when the number of moles of each element on the left side of the equation is equal to the number of moles of each element on the right side of the equation, the chemical equation is said to be **balanced**. In the equation above, there are 2 nitrogen and 6 hydrogen atoms on each side of the equation. Nitrogen is diatomic, which means that two moles of nitrogen atoms are required for this reaction. Next, hydrogen is also diatomic, and you can see from the coefficient that three moles of diatomic hydrogen are required for this reaction—that means 6 moles of hydrogen atoms. On the right side of the reaction, the coefficient of 2 tells us that there are two moles of nitrogen atoms, and 2 × 3 or 6 moles of hydrogen, and the equation is balanced.

Generally speaking, balancing equations is a trial and error process, but it's usually best to use the compound that has the most constituents as your starting point. Let's look at the following equation:

$$HCl + MnO_2 \rightarrow H_2O + MnCl_2 + Cl_2$$

In this equation, there are two reactants (HCl and MnO_2) and three products (H_2O, $MnCl_2$, and Cl_2). To balance it, let's start with MnO_2. Since one molecule of MnO_2 contains two oxygen atoms, at least two molecules of water (H_2O)

must be produced in order to give us two oxygen atoms on the right side of the equation. Therefore,

- Put a two in front of H_2O

Next we notice that a total of 4 chlorine (Cl) atoms appear on the right hand side of the equation, so we must start with four moles of HCl molecules.

- Put a 4 in front of HCl

Sometimes, as you balance the more complex molecules, the rest of the equation becomes balanced. Our equation is now:

$$4\ HCl + MnO_2 \rightarrow 2\ H_2O + MnCl_2 + Cl_2$$

and it's balanced.

STOICHIOMETRY OF A CHEMICAL EQUATION

The stoichiometry of a chemical equation is the calculation of the quantities of reactants and products involved in the reaction. If you're given the amount of a reactant or product (usually in grams), you can calculate the amount of the other products and reactants.

Take a look at our previously balanced equation:

$$4\ HCl + MnO_2 \rightarrow 2\ H_2O + MnCl_2 + Cl_2$$

What if you were asked to calculate the gram quantity of water molecules produced from 10g of magnesium dioxide (MnO_2), according to the chemical equation above? Here are the steps you should take:

Step 1: Convert the gram quantity of MnO_2 into number of moles (Molecular weight = 86.9 g/mol)

$$10g\ MnO_2 \left(\frac{1\ mol}{86.9g} \right) = 0.115\ mol\ MnO_2$$

Step 2: Using the balanced equation, calculate the mole quantity of H_2O

$$0.115\ mol\ MnO_2 \left(\frac{2\ mol\ H_2O}{1\ mol\ MnO_2} \right) = 0.230\ moles\ H_2O$$

Step 3: Convert the number of moles of H_2O into a gram quantity (Molecular weight = 18.0 g/mol)

$$0.230\ moles\ H_2O \left(\frac{18.0g}{1\ mol} \right) = 4.14g\ of\ H_2O$$

If you're given the gram quantity of two reactants, you will need to determine which one is the limiting reagent in order to calculate the amount of product yielded by the reaction. The limiting reagent is the reactant that will be entirely consumed first. Look back at the previous equation. If we started with 10 grams of MnO_2 and 10 grams of HCl, which reactant would be our limiting reagent?

Step 1: First convert the gram quantities of both reactants into moles (MnO_2 MW = 86.9 g/mol and HCl MW = 36.5g/mol)

$$10g\ MnO_2 \left(\frac{1\ mol}{86.9g}\right) = 0.115\ mol\ MnO_2$$

$$10g\ HCl \left(\frac{1\ mol}{36.5g}\right) = 0.274\ mol\ HCl$$

Step 2: Determine the limited reagent using the molar ratios from the chemical equation.

We know that four moles of HCl are used for every one mole of MnO_2. So if we start with 0.274 moles of HCl, then only 0.068 moles of MnO_2 will be used in the reaction—you can set up a proportion to figure this out:

$$.274\ mol\ HCl \left(\frac{1\ mol\ MnO_2}{4\ mol\ HCl}\right) = 0.068\ mol\ MnO_2$$

Therefore, the HCl will be used up first and is the limiting reagent.

It is important to use a balanced chemical equation with its molar ratios to calculate the limiting reagent. But keep in mind that the reactant with the least number of moles should not be automatically labeled as the limiting reagent! Remember that the limiting reagent is the reactant that will be *depleted* first, and it is necessary to use the molar ratios of the reactants to determine what proportions are needed.

STOICHIOMETRY

1. What is the empirical formula of a compound that contains 36% calcium and 64% chlorine by mass?

   A. CaCl

   B. Ca_2Cl

   C. $CaCl_2$

   D. Ca_2Cl_4

   Answer:

   C. First calculate the quantity of each element in a 100 gram sample:

   $$36\% \text{ of } 100 \text{ g} = 36 \text{ g Ca}$$

   $$64\% \text{ of } 100 \text{ g} = 64 \text{ g Cl}$$

   Next convert grams into moles:

   $$36 \text{ g Ca} \times 1 \text{ mol}/40 \text{ g} = 0.9 \text{ mol Ca}$$

   $$64 \text{ g Cl} \times 1 \text{ mol}/35 \text{ g} = 1.8 \text{ mol Cl}$$

   Next set up the ratio:

   $$Ca:Cl = 0.9:1.8 = 1:2$$

   Therefore, the empirical formula is $CaCl_2$.

2. $CaCO_3 \ (s) \rightarrow CaO \ (s) + CO_2 \ (g)$

   The equation above reveals how calcium carbonate can be decomposed into CaO and CO_2. If 75 grams of $CaCO_3(s)$ are initially decomposed, how many grams of CaO are produced?

   A. 21 g

   B. 42 g

   C. 75 g

   D. 84 g

Answer:

B. First convert grams of $CaCO_3$ into moles.

$$FW\ CaCO_3 = 40 + 12 + 3(16) = 100\ g$$

$$75\ g\ CaCO_3 \times 1\ mol/100g = 0.75\ mol\ CaCO_3$$

Next calculate moles of CaO (s):

$$0.75\ mol\ CaCO_3 \times 1\ mol\ CaO/1\ mol\ CaCO_3 = 0.75\ mol\ CaO$$

Finally convert to grams of CaO:

$$0.75\ mol \times 56\ g\ /\ 1\ mol = 42\ g\ CaO$$

3. $CaO(s) + 3C(s) \rightarrow CaC_2(s) + CO(g)$

   Calcium oxide is reacted with solid carbon to form CaC_2 (s) and CO (g). If 72 g of solid C(s) are initially reacted how many grams of $CaCl_2$ are produced?

   A. 3.5 g
   B. 32 g
   C. 64 g
   D. 128 g

Answer:

D. First convert grams of C into moles:

$$72\ grams\ C \times 1\ mol/12\ g = 6\ moles\ C$$

Next calculate moles of CaC_2:

$$6\ moles\ C \times 1mol\ CaC_2/3\ mol\ C = 2\ moles\ CaC_2$$

Finally convert into grams of CaC_2:

$$2\ moles\ CaC_2 \times 64g\ /\ 1\ mol = 128\ g\ CaC_2$$

4. $C_3H_8 + 5O_2 \rightarrow 4H_2O + 3CO_2$

   According to the reaction above, propane reacts with oxygen to form water and carbon dioxide. If 132 g of C_3H_8 are to be reacted with 384 g of O_2, which reactant would be the limiting reagent?

A. C_3H_8

B. O_2

C. H_2O

D. CO_2

Answer:

B. The limiting reagent is the reactant that will run out first during the reaction. First cross off answer choices C and D since they are products and not reactants.

Next convert grams into moles:

$$132 \text{ g } C_3H_8 \times 1 \text{ mol}/44 \text{ g} = 3 \text{ mol } C_3H_8$$

$$384 \text{ g } O_2 \times 1 \text{ mol}/32 \text{ g} = 12 \text{ mol } O_2$$

Using one reactant, determine how many moles of the other reactant would be required if we started with the quantity calculated above:

Considering 3 mol of C_3H_8:

3 mol $C_3H_8 \times$ 5 mol O_2/1 mol C_3H_8 = 15 moles O_2 would be required

Starting the reaction with 384 g of O_2, we only calculated that we had 12 moles of O_2. Therefore, the limiting reagent is therefore O_2.

5. What is the empirical formula of a compound comprised of 63% Mn and 37% O?

A. MnO

B. Mn_2O

C. MnO_2

D. MnO_3

Answer:

C. First calculate the quantity of each element in a 100 g sample:

$$63\% \text{ Mn} = 63 \text{ g Mn}$$

$$37\% \text{ O} = 37 \text{ g O}$$

Next calculate the moles of each:

$$63 \text{ g Mn} \times 1 \text{ mol}/55 \text{ g} = 1.15 \text{ mol}$$

$$37 \text{ g O} \times 1 \text{ mol}/16 \text{ g} = 2.3 \text{ mol}$$

Next calculate the mole to mole ratio between elements:

$$\text{Mn:O} = 1.15 : 2.3 = 1 : 2$$

Therefore, the empirical formula is MnO_2.

Chapter 6

OXIDATION-REDUCTION REACTIONS

OXIDATION-REDUCTION REACTIONS

An atom's **oxidation state** (or **oxidation number**) is the charge that the atom would carry in a bond if all of the participating electrons belonged to more electronegative atoms. An atom only exhibits an oxidation state when it's bonded to other atoms. If a substance is comprised of monoatomic ions like NaCl, in which the electrons are completely acquired by the more electronegative atom (chlorine), then each ion's oxidation number equals the ionic charge. Sodium's oxidation number in NaCl is +1, while chlorine's oxidation number is –1. Sodium chloride is a neutral molecule, which means the oxidation numbers add up to zero. Polyatomic ions such as MnO_4^- have an overall charge, and the sum of the oxidation numbers of their individual atoms will add up to the overall charge of the ion. The oxidation state of oxygen in MnO_4^- is –2. Since there are four oxygen atoms, each with a charge of –2, the oxidation state of manganese (Mn) will be +7 in order to yield an overall charge of –1.

If atoms are involved in covalent bonds, as they are in HCl, the oxidation state designates a theoretical charge. Since the shared electron pair is hanging around the electronegative chlorine atom most often, a theoretical charge of –1 can be assigned to the chlorine atom and a charge of +1 can be assigned to hydrogen. When atoms of the same element are bonded together as they are in O_2, the electron pair is shared equally and the oxidation number of each atom is zero.

The oxidation state of an atom depends on the other atoms to which it is bonded, meaning that some atoms have different oxidation numbers when they're involved in different molecules. For example, the oxidation state of manganese in MnO_4^- is +7, while it's +4 in MnO_2. Generally speaking, you can use the following rules when assigning oxidation numbers:

1. The oxidation number of atoms in an elementary compound are zero. (Examples: Cl_2, O_2, F_2, N_2, and P_4)
2. The oxidation number of an atom in Group 1A (the alkali metals) is generally +1; the oxidation number of an atom in Group 2A (the alkaline earth metals) is generally +2; the oxidation number of an atom in Group 3A is generally +3.
3. The oxidation number of fluorine (F) is usually –1.
4. The oxidation number of oxygen (O) in a molecule is usually –2 (except in H_2O_2 as when bonded to any metal; then its oxidation number is –1).
5. The oxidation number of hydrogen (H) in a molecule is usually +1.
6. The oxidation number of any atom in Group 8A (noble gases) is 0.
7. The sum of all the oxidation numbers in a neutral molecule is equal to zero. The sum of all the oxidation numbers in a polyatomic ion is equal to the overall charge of the ion.

To determine an atom's oxidation number in a neutral molecule, simply apply the above rules and then solve for the remaining atom, if you need to.

Some of the transition elements have oxidation states that differ, and the different oxidation states of these transition elements are denoted by Roman numerals. For example, iron (Fe) in $FeSO_4$ has an oxidation number of +2. These ions are considered to be iron (II) ions. But iron (Fe) in Fe_2O_3 has an oxidation number of +3, and these ions are identified as Fe (III).

It is a good idea to also familiarize yourself with the oxidation states of some common polyatomic ions:

ammonium	NH_4^+
carbonate	CO_3^{2-}
chlorate	ClO_3^-
cyanide	CN^-
hydrogen carbonate (bicarbonate)	HCO_3^-
hydrogen sulfate (bisulfate)	HSO_4^-
hydroxide	OH^-
nitrate	NO_2^-
nitrate	NO_3^-
perchlorate	ClO_4^-
permanganate	MnO_4^-
phosphate	PO_4^{3-}
sulfate	SO_4^{2-}
sulfite	SO_3^{2-}

REDOX REACTIONS

Oxidation numbers generally change during the course of chemical reactions. Let's look at the following oxidation-reduction reactions, also known as redox reactions. A **redox reaction** is a reaction in which electrons are transferred between substances, resulting in a change in at least one element's oxidation state.

When an atom *gains* electrons, its oxidation number decreases (the atom becomes more negative). The atom that gains electrons is said to have been **reduced**. When an atom *loses* electrons, its oxidation number increases (the atom becomes more positive). The atom that loses electrons is said to have been **oxidized**. The following mnemonic may help you remember these terms:

 LEO the lion says GER
 LEO: Lose Electrons during Oxidation
 GER: Gain Electrons during Reduction

Oxidation does not occur without reduction. Think of it this way: if one atom is gaining electrons (being reduced), then there has to be another atom that's losing electrons (being oxidized). Remember, the total number of electrons involved in a chemical equation will not change.

A **reducing agent** (or reductant) is an atom that's being oxidized and causing another atom to be reduced. An **oxidizing agent (or oxidant)** is an atom that is being reduced and causing another atom to be oxidized.

Let's take a look at the following reaction:

$$Fe + 2HCl \rightarrow FeCl_2 + H_2$$

In order to determine how the electrons are redistributed during the course of this reaction, start by writing the oxidation numbers above each atom in the chemical equation:

$$\overset{0}{Fe} + 2\overset{+1\ -1}{HCl} \rightarrow \overset{+2\ -1}{FeCl_2} + \overset{0}{H_2}$$

From this equation, we can tell the following:

- Iron loses two electrons—its oxidation number changes from 0 to +2.

- Each of the two hydrogen atoms gains an electron—their oxidation numbers change from +1 to 0.

- Chlorine's oxidation state remains the same.

Since iron has lost electrons, it has been oxidized and we know that it's the reducing agent. Each hydrogen atom has been reduced—in this case, hydrogen is the oxidizing agent.

The exchange of electrons in a reaction can be depicted through **half reactions**:

$$\text{oxidation: } Fe \rightarrow Fe^{2+} + 2\ e^-$$

$$\text{reduction: } 2\ H^+ + 2\ e^- \rightarrow H_2$$

On the chemistry GSE, you might be asked to balance an oxidation-reduction reaction. You'll probably find that writing out the half reactions will be very helpful for balancing more complicated reactions.

Let's look at the following unbalanced equation:

$$Cu + HNO_3 \rightarrow Cu(NO_3)_2 + H_2O + NO$$

In order to balance this equation, we first write out the oxidation states for each element or polyatomic ion:

$$\overset{0}{Cu} + \overset{+1}{H}\overset{+5-2}{(NO_3)} \rightarrow \overset{+2}{Cu}\overset{+5-2}{(NO_3)_2} + \overset{+1-2}{H_2O} + \overset{+2-2}{NO}$$

Next we write out the half reactions:

oxidation: $Cu^0 \rightarrow Cu^{2+} + 2e^-$

reduction: $N^{+5}O_3 + 3e^- \rightarrow N^{+2}O^{-2}$

As you know, the total number of electrons lost during oxidation must equal the total number of electrons gained during reduction. In order to balance the electrons in this reaction, first find their lowest common multiple. You probably know that the lowest common multiple of two and three is six. The equations will therefore be balanced as follows:

oxidation: $[Cu^0 \rightarrow Cu^{2+} + 2e^-] \times 3$

reduction: $[N^{+5}O_3 + 3e^- \rightarrow N^{+2}O^{-2}] \times 2$

oxidation: $3Cu^0 \rightarrow 3Cu^{2+} + 6e^-$

reduction: $2(N^{+5}O_3) + 6e^- \rightarrow 2(N^{+2}O^{-2})$

Put the two equations together and you have the following:

$$3Cu + 2(NO_3) \rightarrow 3Cu^{2+} + 2(N^{+2}O^{-2})$$

Bring back the original unbalanced equation and fill in the numbers accordingly:

$$3Cu + 2HNO_3 \rightarrow 3Cu(NO_3)_2 + H_2O + 2NO$$

Is it balanced yet? If we count the number of oxygen atoms, we see that there are 21 oxygen atoms on the right side of the equation and only 6 oxygen atoms on the left side. Therefore, we need to increase the coefficient of HNO_3. If we increase the coefficient to seven, we can't balance the seven hydrogen atoms that we get. Therefore, we need to try the next highest *even* number. If we plug an eight into the equation, we then have to increase the number of hydrogen atoms on the right side of the equation. By increasing the coefficient of H_2O, we can balance the entire chemical equation:

$$3Cu + 8HNO_3 \rightarrow 3Cu(NO_3)_2 + 4H_2O + 2NO$$

OXIDATION-REDUCTION REACTIONS

1. The oxidation state of oxygen is equal to –2 in all of the following compounds EXCEPT?

   A. MnO_2

   B. H_2O

   C. O_2

   D. OH^-

   Answer:

   C. Watch out...this question asked for the EXCEPTION. The oxidation state of oxygen in three of the answer choices will be equal to –2. Start with answer choice A and determine the oxidation state. Stop once you have found the answer choice where oxygen's oxidation state is not equal to –2.

   The answer is C, the oxidation state of elemental oxygen is equal to 0.

2. Which of the following is a strong reducing agent?

   A. Zn

   B. H^+

   C. Na^+

   D. Cl_2

   Answer:

   A. Zinc is a reducing agent because it likes to give up electrons in an oxidation reaction.

   $$Zn \rightarrow Zn^{2+} + 2e^-$$

   The other answer choices are examples of oxidizing agents that like to accept electrons in reduction reactions.

   $$2H^+ + 2e^- \rightarrow H_2$$

   $$Na^+ + e^- \rightarrow Na$$

   $$Cl_2 + 2e^- \rightarrow 2Cl^-$$

3. According to the following reaction how many moles of electrons are transferred:

$$MnO_4^- (aq) + 8H^+ + 5Fe^{2+} (aq) \rightarrow Mn^{2+} (aq) + 4H_2O + 5Fe^{3+} (aq)$$

A. 2

B. 3

C. 4

D. 5

Answer:

D. First determine the oxidation state of the elements in each compound:

$$Mn^{7+}O_4^- + 5Fe^{2+} \rightarrow Mn^{2+} + Fe^{3+}$$

Next write out the half reactions:

$$Mn^{7+}O_4^- + 5e^- \rightarrow Mn^{2+}$$

$$5 (Fe^{2+} \rightarrow Fe^{3+} + e^-)$$

The total number of electrons transferred is equal to 5.

4. If an iron nail is placed in a copper (II) sulfate solution, solid copper metal will plate on the surface of the nail. Which of the following has occurred?

A. Cu (II) is reduced to Cu (s)

B. Cu (II) is oxidized to Cu (s)

C. Fe (s) is reduced to Fe^{2+}

D. Both Cu (II) and Fe(s) are oxidized

Answer:

A. The question tells us that we are starting with copper (II) sulfate solution:

$$Cu^{2+}SO_4^{2-}$$

If solid copper is plating on the surface of the nail then the following reaction is occurring:

$$Cu^{2+} + 2e^- \rightarrow Cu(s)$$

Since electrons are being gained, the copper is reduced to its solid form. Copper acquires the electrons from the iron which is being oxidized: $Fe(s) \rightarrow Fe^{2+} + 2e^-$

5. What is the oxidation number for chlorine in potassium dichromate, $K_2Cr_2O_7$?

A. −14

B. −2

C. +2

D. +6

Answer:

D. According to the general oxidation rules, the oxidation number for oxygen is generally −2 (since an oxygen atom usually acquires two electrons when bonded). The oxidation number for an atom in group IA is generally +1 (since alkali metals usually lose their lone valence electron when bonded). Using these rules we can determine the oxidation number for chromium in $K_2Cr_2O_7$ as follows:

$$+1 \quad -2$$

$$K_2Cr_2O_7$$

$$xK_2 + xCr_2 + xO_7 = 0$$

$$xCr_2 = 0 - xK_2 + xO_7$$

$$xCr_2 = 0 - (+2) - (-14)$$

$$xCr_2 = +12$$

The oxidation number for chromium is therefore:

$$xCr = +12/2 = +6$$

Chapter 7

GASES AND PHASE PHANGES

IDEAL GAS LAW

Gases differ from liquids and solids in part because they are easy to compress into smaller volumes. An ideal gas obeys the gas laws over all temperatures and pressures and can be defined by the ideal gas law equation. The pressure, temperature, volume, and molar quantity of an ideal gas are related in the following way:

$$PV = nRT$$

P = pressure (atm)

V = volume (L)

n = molar quantity (mol)

T = temperature (K)

R = gas constant ($0.0821 \frac{L - atm}{mol - K}$)

From this equation, you can see that pressure and volume are inversely proportional. This means that as pressure increases, volume decreases and vice versa, assuming a constant temperature. Temperature is directly proportional with both volume and pressure, so if the pressure is held constant as temperature increases, volume increases; as temperature decreases, so does volume. As temperature increases, so does pressure, and vice versa (assuming a constant volume).

COMBINED GAS LAW

The following manipulation of the ideal gas equation is called the **combined gas law**, and depicts the relationship between pressure, temperature, and volume.

$$\frac{P_1 V_1}{T_1} = \frac{P_2 V_2}{T_2}$$

Temperature is usually measured in degrees Celsius (°C) or degrees Kelvin (K), but you must make sure that when you are doing calculations using any of the gas laws, you are not using two different units, either for temperature or pressure! In addition to this, the temperature *must* be in Kelvins when you're using the ideal gas law. Use the following conversion to get temperatures in the same units:

Kelvin (K) = degrees Celsius (°C) + 273

Zero degrees Kelvin (–273 °C) represents absolute zero—the lowest possible temperature.

Pressure is the measure of force per unit area of surface $\left(P = \frac{F}{A}\right)$. It can be measured in Pascal (Pa), atmospheres (atm), or millimeters of mercury (mmHg or torr). Millimeters of mercury refers to the mercury that rises in a barometer, an instrument used for measuring pressure. Torr and mmHg are the same thing.

1 atm = 760 mmHg or torr = 1.013×10^5 Pa

Standard temperature and pressure (STP) is 0 °C (273K) and 1 atm. At STP, 1 mole of gas occupies 22.4 L. You should remember this for the exam.

$$PV = nRT$$

$$V = \frac{nRT}{P} = \frac{(1 mol)(0.0821 \text{ L} - \text{atm} / \text{mol} - \text{K})(273 \text{K})}{1 \text{ atm}} = 22.1 \text{L}$$

The kinetic energy of a gas is directly proportional to its temperature. Gases at higher temperatures have greater kinetic energy than those at lower temperatures.

$$KE = \frac{3}{2} nRT$$

KE = kinetic energy (joules, J)

T = temperature (K)

n = molar quantity

R = gas constant (8.31 J/mol-K)

Notice that the R value in this equation is different from the one in the ideal gas equation. Since kinetic energy is measured in joules, the gas constant must be in J/mol-K.

The kinetic energy of a gas molecule is also related to the molecule's mass and velocity.

$$KE = \frac{1}{2}mv^2$$

KE = kinetic energy (J)

m = mass (kg)

v = velocity (m/s)

You can assume that all gases you'll see on the exam are ideal and generally adhere to the following guidelines:

- The volume of the gas molecules is significantly smaller than the total volume of space in which the molecules are moving around.
- There are no attractive forces between gas molecules. All collisions that occur between molecules are perfectly elastic, with no loss of kinetic energy.
- The collisions between gas molecules occur very quickly, and the total amount of time during collisions is much smaller than the time between collisions.

The ideal gas laws begin to break down at extremely low temperatures and high pressures. Gases that are at low temperature or high pressure are real gases and generally exhibit the following properties:

- Gas molecules will stick together upon colliding, as a result of the van der Waals attractions between the molecules.

- The volume of the gas molecules makes up a large proportion of the total volume of space in which they move.

When we measure the volume of ideal gases, we are measuring the amount of empty space that exists around the gas molecules. The value is generally pretty close to the total volume of space that the gas molecules occupy since the molecules themselves generally take up minimal space. But under real conditions, the total volume that the gas molecules occupy is a significant part of the total volume, so the calculated volume is less than the actual.

Pressure represents the force that the gas molecules exert when they bump into the walls of their container. Under real conditions, the gas molecules stick together because of van der Waals forces; this minimizes the number of collisions that occur. This means that the pressure exerted by ideal gases is greater than that exerted by real gases.

DALTON'S LAW OF PARTIAL PRESSURE

Several different gases within the same container will have the same volume, temperature, and pressure. In a mixture of several different gases, the total pressure is equal to the sum of the partial pressures of each gas. This is known as **Dalton's law**:

$$P_{total} = P_{gas1} + P_{gas2} + P_{gas3} + \ldots$$

The partial pressure of each gas is dependent on the number of moles of that particular gas present in the mixture. A gas that makes up 30% of a mixture will have a partial pressure that is 30% of the total pressure.

GRAHAM'S LAW

Earlier we mentioned that a gas molecule's kinetic energy is directly proportional to its temperature. Therefore, we can assume that all gases in a mixture at the same temperature have the same average kinetic energy. We also mentioned that the kinetic energy of a gas molecule is dependent on both its mass and velocity:

$$KE = \frac{1}{2}mv^2$$

If we have a mixture of hydrogen gas and oxygen gas in a container, the kinetic energy of the two gases would be the same. Since oxygen molecules are heavier than hydrogen molecules, the velocity of the oxygen molecules would be less than that of the hydrogen molecules, according to the kinetic energy equation. This concept is known as **Graham's law**, which basically says that the rates of effusion (or leaking out through a small opening) of gases are inversely proportional to the square roots of their densities at constant temperature and pressure.

$$\frac{\text{effusion rate (gas A)}}{\text{effusion rate (gas A)}} = \sqrt{\frac{\partial_{gasB}}{\partial_{gasA}}}$$

PHASE CHANGES

What happens when you put a pot of water on the stove and turn on the burner underneath it? The temperature of the water increases, and so does its kinetic energy, until the water reaches a temperature of 100 °C. When it reaches 100 °C, the kinetic energy of the molecules is great enough to break the hydrogen bonds that hold the water together. As additional heat is added to the water, the kinetic energy will not increase—the newly inputted energy goes toward breaking intermolecular forces. At this stage, water will undergo a phase change from liquid to gas. Since there is no change in kinetic energy during phase changes, there is no increase in temperature, and the water will remain at 100 °C as it undergoes transformation from liquid to gas. If additional heat is added after the phase change is complete, the temperature and kinetic energy of the gaseous water (steam) will again increase proportionally.

The amount of heat that a substance must absorb in order to change from a liquid to a gas is known as its heat of vaporization. This value varies for different substances. For instance, the heat of vaporization of water is 540 cal/g. You will not need to memorize the heat of vaporization of any substance for the chemistry GSE.

If we started with solid water (ice) and added heat, the temperature and kinetic energy of the ice molecules would increase until the substance reached 0 °C. At that point, the temperature of the molecules (and their overall kinetic energy) would stay the same until the ice was completely converted to liquid water. The amount of heat that's required to drive a substance from solid to liquid phase is known as the heat of fusion. In the case of water, the heat of fusion is 80 cal/g. As soon as all of the ice has been converted into liquid water, the temperature and kinetic energy of the liquid will again start to increase.

Specific heat refers to the amount of heat required to raise the temperature of 1 g of a substance by 1 °C. Specific heat depends on the identity of the substance, as well as the phase that it's in. For example, the specific heat for liquid water is 1 cal/g-°C, while for steam it's 0.447 cal/g-°C, and for ice it's 0.502 cal/g-°C. Remember that the higher the specific heat, the more heat that's required to raise or lower a substance's temperature.

Not surprisingly, phase changes also occur as heat is removed from a substance. If heat is removed from steam, the temperature and kinetic energy will also decrease at a proportionate rate until it reaches 100 °C. Heat is given off and intermolecular bonds form to create the liquid phase. The phase change will occur again when the temperature of the water hits 0 °C, and heat is given off to form bonds that result in the solid state of water. Remember:

- When a substance changes from liquid to gas, it is being vaporized and the heat of vaporization is absorbed to cause the phase change.

- When a substance changes from gas to liquid, it is being condensed and the heat of vaporization is released to cause the phase change.

- When a substance changes from solid to liquid, it is melting and the heat of fusion is absorbed to cause the phase change.

- When a substance changes from liquid to solid, it is freezing and the heat of fusion is released to cause the phase change.

The stronger the intermolecular bonds in a substance, the greater its heat of fusion or heat of vaporization. For example, it is easier to melt a piece of ice than a piece of copper because the intermolecular bonds in ice are significantly weaker than the ones in copper. For this reason, it requires more energy to break copper's intermolecular bonds, and its heat of fusion is much higher.

Why is water's heat of fusion less than its heat of vaporization? Well, a greater number of intermolecular bonds are broken when water changes from a liquid to a gas than when it changes from solid to liquid. It's also true that more intermolecular bonds are *formed* when water changes from a gas to a liquid than when it changes from a liquid to a solid.

The following phase change diagram depicts the associations between heat and temperature.

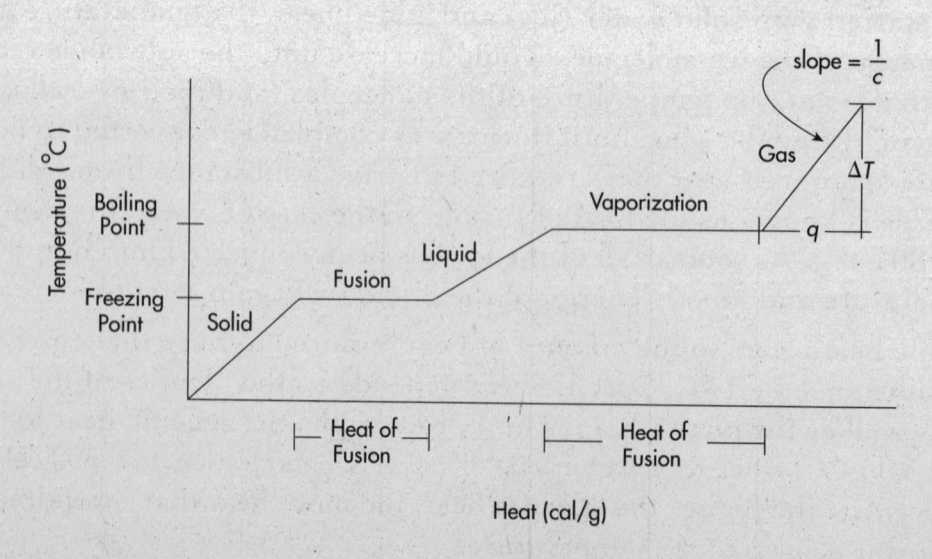

The phase changes are represented by horizontal lines. This graph shows clearly that temperature remains constant as heat is increased during phase changes. The greater the length of the line, the more heat required during the phase change.

The sloped line depicts the relationship between temperature and heat within a phase, and shows that the increase in temperature is directly proportional to the increase in heat. The slope of the line depends on the substance's specific heat; the steeper the line, the lower the specific heat.

You will most likely be required to determine relationships between heat and temperature. Use the following equation to determine the relationship.

$$q = mc\Delta T$$

q = heat added to a substance

m = mass of the substance

c = specific heat for that substance

ΔT = change in temperature of the substance

There is another type of phase change diagram you should be able to interpret.

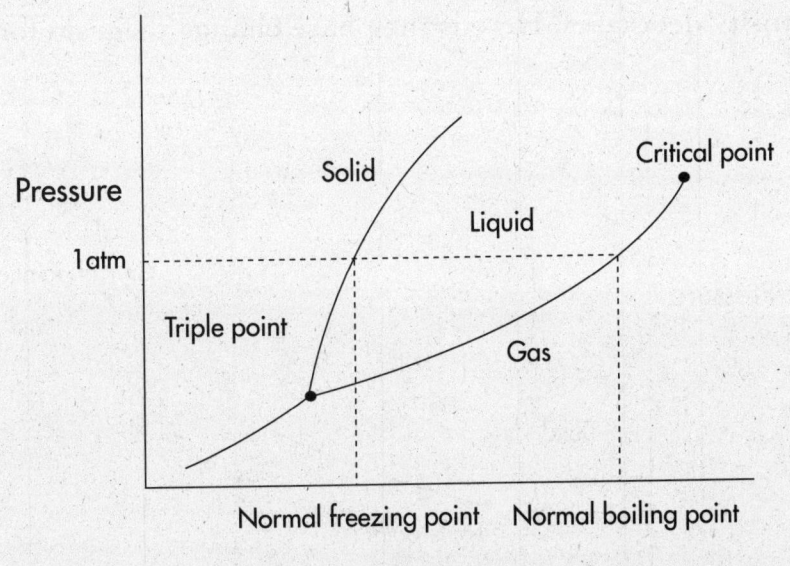

GASES AND PHASE CHANGES ♦ 71

This diagram depicts the relationship between the temperature and pressure of a substance. The three lines represent the three different phases: solid, liquid, and gas. As you cross over each line, the phase of the substance changes, and the line itself represents a time at which the two phases are at equilibrium. The point at which all three phases are at equilibrium is known as the triple point.

As you can see from the diagram, as temperature increases, higher pressure is required in order to move a substance from a gaseous to liquid state. The temperature above which the liquid state no longer exists (regardless of pressure increase) is known as the critical point, or critical temperature.

For most substances, the solid state is denser than the liquid state. This means that the solid state of a substance will not float if it's immersed in its liquid state. One exception to this generality is water. As you probably know, ice floats in a glass of water, and this is because solid water is less dense than liquid water. As water cools, hydrogen bonds are formed between water molecules, which then arrange themselves in a crystal structure. This rigid crystal arrangement actually increases the overall volume of the substance.

As you know, density is a quantity that relates the mass of a substance to its volume:

$$\partial = \frac{m}{v}$$

Volume and density are inversely proportional. If the volume of ice is increased, its density decreases. Here is the phase change diagram for water:

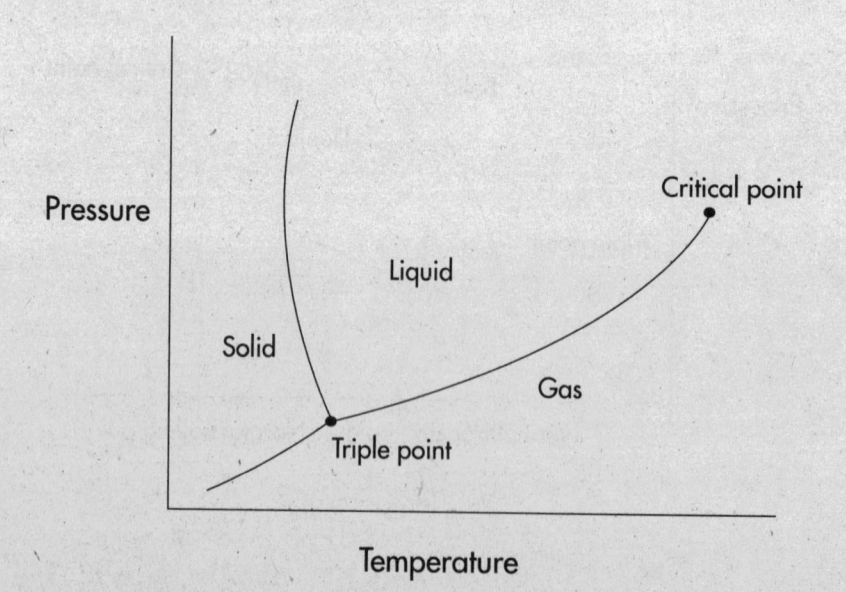

Most substances undergo a phase change from liquid to solid as you increase pressure at a constant temperature. However, the opposite is true of water. As you increase the pressure of ice at a constant temperature, it will melt. This fact is exhibited by ice skating—as the skates' blades run across the solid surface of a lake, the pressure they exert causes a phase change to the liquid state, creating a slippery surface.

At low pressures, some substances can change directly from the solid state to the gaseous state, in a process known as sublimation. Dry ice is an example of a substance that's capable of sublimation.

GASES AND PHASE CHANGES

1. A gas sample is held in a 800 mL container at a temperature of 25°C. If the gas exerts a pressure of 10.0 atm on the walls of its container, how many moles of gas are present in the container? (R = 0.0821 L-atm/mol-K)

A. (0.0821)(25)/(10)(800)

B. (10)(800)/(0.0821)(25)

C. (10)(0.8)/(0.0821)(298)

D. (0.0821)(298)/(10)(0.8)

Answer:

C. According to the ideal gas equation, $PV = nRT$

Therefore,

$$n = \frac{PV}{RT}$$

Be careful…it is necessary to convert terms to the appropriate units.

Temperature: 25°C + 273 = 298K

Volume: 800 mL × 1L/1000 mL = 0.8 L

$$n = \frac{(10)(0.8)}{(0.0821)(298)}$$

2. A particular sample of an ideal gas has a pressure of 10 atm at a temperature 25 °C. If volume is held constant and pressure is increased to 20 atm, how is temperature effected?

A. Temperature stays the same

B. Temperature increases to 50 °C

C. Temperature increases to 100 °C

D. Temperature increases to 323 °C

Answer:

D. According to the ideal gas law $PV = nRT$, pressure and temperature are directly proportional at a constant volume.

$$\frac{P_1}{T_1} = \frac{P_2}{T_2}$$

$$T_2 = \frac{P_2 T_1}{P_1}$$

If pressure is doubled the temperature must also be doubled when held at a constant volume but first temperature must be converted to Kelvin. 25 °C = 298 K, and (298 K)(2) = 596 K. Converted to Celsius again, you get 323 °C. (Conversion is K = °C + 273.)

3. At a given temperature, If a gas mixture contained hydrogen, oxygen, nitrogen and helium, which gas will have the lowest rate of effusion?

A. H_2

B. O_2

C. N_2

D. He

Answer:

B. According to Grahams Law, $\frac{r_1}{r_2} = \frac{\sqrt{M_2}}{\sqrt{M_1}}$, the heavier the gas the lower the rate of effusion. Of the gases within this mixture, O_2 has the largest molecular weight and will therefore experience the lowest rate of effusion.

The next two questions are based on the phase diagram below.

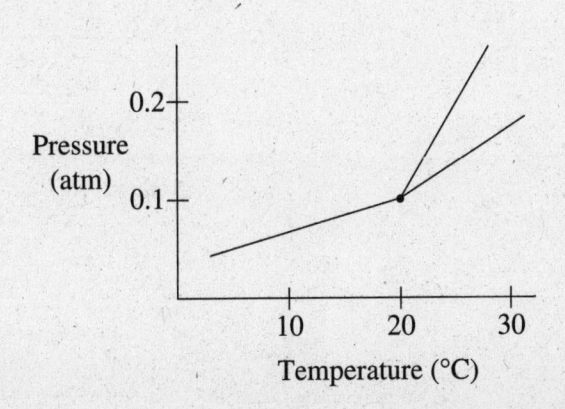

GASES AND PHASE CHANGES ◆ 75

4. At a temperature of 30° C, if the pressure is decreased from 0.2 atm to 0.1 atm, which phase change occurs?

A. Melting

B. Freezing

C. Vaporization

D. Condensation

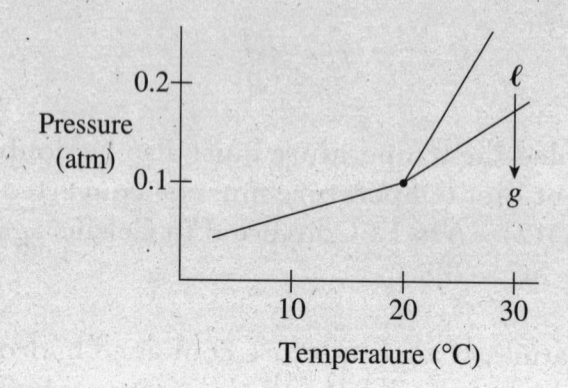

Answer:

C. Looking at the phase diagram, if pressure is decreased from 0.2 atm to 0.1 atm, the substance is changing from the liquid phase to the gas phase, which is known as vaporization.

5. If the temperature of the substance illustrated in the above diagram is increased, which of the following phase changes can occur?

   I. Melting

   II. Vaporization

   III. Sublimation

A. I only

B. I and II only

C. II only

D. I, II and III

Answer:

D. Insert Phase Diagram B

As we increase temperature the following phase changes could occur as depicted above:

$$\text{Solid} \rightarrow \text{Liquid (Melting)}$$

$$\text{Liquid} \rightarrow \text{Gas (Vaporization)}$$

$$\text{Solid} \rightarrow \text{Gas (Sublimation)}$$

Chapter 8

SOLUTIONS

A solution is a homogeneous mixture in which one or more substances are uniformly dispersed as ions, atoms, or molecules. The majority of chemical reactions occur in solution.

Solubility is the amount of solute that can dissolve in a given solvent to form a saturated solution; solubility varies with temperature and pressure. Think of a sample of sodium chloride (NaCl) dissolved in water. The mixture of sodium and chlorine ions in water is an example of a solution. The water is known as the solvent, since it is the component in the greatest quantity and is responsible for dissolving the salt. The salt is the solute—the solitude is basically any solution component that isn't the solvent. This saltwater mixture, and any other mixture in which water is the solvent, are known as aqueous solutions.

A solution becomes more concentrated as more solute is added to the solvent. A weak solution consists of relatively small quantities of solute and is also called a dilute solution, whereas concentrated solutions are ones in which lots of solute is dissolved in the solvent. Solution concentration is determined and recorded by a measurement called molarity.

MOLARITY

Molarity (*M*) is the most common unit of concentration in chemistry; it is a measurement of the number of moles of solute dissolved per total liters of solution. Brackets are used to denote molar concentration. For example, [HCl] represents the concentration (in moles/L) of hydrogen chloride.

$$\text{molarity}(M) = \frac{\text{moles of solute}}{\text{liters of solution}}$$

For example, the molarity of a solution made up of 0.60 moles of fructose ($C_6H_{12}O_6$) dissolved in 3.0 liters of water is:

$$\frac{.60 \text{ mol fructose}}{3.0 \text{ L solution}} = 0.20 \, M$$

PERCENT BY MASS

Concentrations of solutions can also be measured in terms of the mass percent of solute.

$$\% \text{ by mass} = \frac{\text{mass of solute}}{\text{mass of solution}} \times 100$$

If we said that a solution was 4.0% solute, that would mean that 4.0 g of solute were dissolved in enough solvent to yield 100 g of solution (96 g).

MOLALITY

Molality (*m*) is another measure of the concentration of a solution:

$$\text{molality}(m) = \frac{\text{mass of solute}}{\text{kilograms of solvent}}$$

For example, the molality of 0.75 mol of sodium chloride dissolved in 0.96 kg of water is:

$$\frac{.075 \text{ mol sodium chloride}}{0.96 \text{ kg water}} = 0.78 \, m$$

MOLE FRACTION

Mole fraction (X) is yet another way of expressing the concentration of a substance. The mole fraction of substance A is equal to the number of moles of A, divided by the total number of moles in the solution.

$$\text{mole fraction}(X_A) = \frac{\text{moles of A}}{\text{total moles of solution}}$$

SATURATION

A saturated solution is one in which no more solute can be dissolved, and saturated solutions are at equilibrium. Earlier we mentioned that the solubility of a substance varies with temperature and pressure. Generally speaking, as you increase the temperature of a solution, the solubility of the solute also increases if a solute is a solid.

SOLUBILITY

A solute that is readily dissolved by a solvent is said to be soluble in that solvent. But how do you know whether or not a solute will be soluble in a particular solvent? The simple rule to memorize here is that like dissolves like. This means that polar or ionic solutes will dissolve in polar solvents. For instance, sodium chloride is an ionic substance that dissolves readily in water, which is a polar solvent. When ionic molecules break up in solution (when they dissociate), the free ions enable the solution to conduct a current. Solutes that provide ions in aqueous solutions are called electrolytes.

Substances that are nonpolar usually dissolve in nonpolar solvents. Often, in order to predict the outcome of a chemical reaction, it is helpful to know about the solubility of the substances involved.

Let's talk about ionic substances. First of all, cations are generally soluble in water. Some examples of soluble cations include the alkali metals (Li^+, Na^+, K^+, Rb^+, and Cs^+) and ammonium (NH_4^+). The solubility of the alkaline earth metals and transition metals depends on the anion to which they're bonded. Anions are also generally soluble. Some examples of these soluble anions are nitrate (NO_3^-), chlorate (ClO_3^-), perchlorate (ClO_4^-) and acetate ($C_2H_3O_2^-$) salts. Chlorine (Cl^-), bromine (Br^-) and iodine (I^-) salts are soluble, except when they are bonded with silver (Ag^+), lead (Pb^+) or mercury (Hg_2^{2+}). Sulfate salts are soluble except when they're bonded with silver (Ag^+), lead (Pb^+), mercury (Hg_2^{2+}), calcium (Ca^{2+}), strontium (Sr^{2+}) and barium (Ba^{2+}).

The following anions are generally insoluble: hydroxide (OH^-), carbonate (CO_3^{2-}), phosphate (PO_4^{3-}), sulfite (SO_3^{2-}), chromate (CrO_4^{2-}), and sulfide (S^{2-}).

COLLIGATIVE PROPERTIES

Colligative properties are ones that rely on the concentration of solute particles in solution and not on the solute's identity. Some examples of colligative properties are vapor pressure, freezing point depression, boiling point elevation, and osmotic pressure.

Vapor Pressure

When a liquid evaporates, the molecules that enter the gas phase exert a pressure called vapor pressure. Not surprisingly, if a solvent contains strong intermolecular bonds, fewer molecules will enter the gas phase and the vapor pressure will be low (the boiling point will be high). Liquids boil when the vapor pressure of the liquid equals the pressure of the surrounding atmosphere. You should also know that when a solute is added to a solvent, the vapor pressure of the resulting solution is lower than that of the pure solvent. Since solute is taking up space at the surface of the liquid, less solvent is able to evaporate, and the vapor pressure decreases.

A couple of factors that determine a liquid's vapor pressure are its temperature and its chemical composition. As we mentioned, the higher the temperature, the higher the vapor pressure of a liquid, and impure solvents that contain one or more solutes have lower vapor pressure than they do in their pure form. Factors that do not affect vapor pressure include the surface area of the liquid and the volume of the liquid in the flask.

A liquid that has high vapor pressure is volatile, which means that it evaporates relatively quickly due to its weak intermolecular bonds. An example of a volatile solution is rubbing alcohol.

Freezing Point Depression and Boiling Point Elevation

Certain solutes, when added to a solvent, act to raise the solvent's boiling point and lower its freezing point. For instance, adding ethylene glycol (CH_2OHCH_2OH) to water decreases water's freezing point to less than 0 °C. In fact, ethylene glycol is the main ingredient in automobile antifreeze because of its effect on water's freezing point. The following formula can be used to calculate the magnitude of freezing point depression:

$$\Delta T_f = k_f m \times i$$

ΔT = change in temperature (freezing point depression)

k_f = freezing point depression constant

m = molality

Boiling point elevation can be measured by the following equation.

$$\Delta T_b = k_b m \times i$$

ΔT = change in temperature (boiling point elevation)

k_b = boiling point elevation constant

m = molality

Osmotic Pressure

A semipermeable membrane will allow solvent particles to travel through it while preventing the passage of solute particles. This phenomenon is known as **osmosis**. The solvent passes through the membrane until its concentration is equal on both sides of the membrane. If two solutions made with the same solvent are separated by a semipermeable membrane, the solvent will travel from the side in which its concentration is higher to the side in which it has a lower concentration, until the concentration of both sides are equal. The amount of pressure that's required to prevent osmosis from occurring across the membrane when one of the liquids is pure solvent is known as osmotic pressure. The more solute that is present in the solution side, the greater the osmotic pressure. Osmotic pressure is therefore a colligative property that's directly related to the molar concentration of solute within a solution. The following equation can be used to calculate osmotic pressure:

$$\Pi V = nRT$$

$$\Pi = MRT \times i$$

Π = osmotic pressure (atm)

n = number of moles of solute

R = the gas constant (0.0821 L-atm/mol-K)

T = temperature (K)

V = total volume of solution (L)

M = molarity of the solution (mol/L)

DENSITY

As we mentioned earlier, the density of a substance is a measure of its mass per unit volume. It is an important property of solids, liquids, and gases, and is expressed by the following equation.

$$\partial = \frac{m}{v}$$

∂ = density (g/L)

m = mass of the substance (solid, liquid, or gas)

V = volume of the substance

Chapter 9

KINETICS AND EQUILIBRIUM

In chemistry, kinetics deals with the rates of chemical processes and how various conditions affect them. The reaction rate refers to the amount of product formed (or reactant consumed) over time. The slowest step of the reaction determines its overall rate. For this reason, the slowest step is known as the rate determining step.

Sometimes a catalyst is added to a reaction in order to speed the reaction up. Although catalysts are added in the same way as are reactants, they are not consumed in the course of the reaction. The chemical makeup of a catalyst may be temporarily changed during the chemical reaction, but by the end of the reaction, the catalyst is back in its original form. Catalysts are usually written above the reaction arrow:

$$2H_2O_2\,(aq) \xrightarrow{HBr} 2H_2O(l) + O_2\,(g)$$

Hydrogen bromide (HBr) is the catalyst in the above chemical reaction.

RATE LAWS

A rate law is an equation that expresses the rate of a reaction in relation to the concentrations of the reactants. At times, the concentration of one reactant will have a significant effect on the overall rate of a reaction, while the concentrations of other reactants will have no effect on the rate at all. All of the reactants are still required in order for the reaction to take place, but having excess of one reactant will not increase the reaction's speed.

For the following reaction:

$$A + B \rightarrow C + D$$

The rate law is:

$$\text{rate} = k[A]^x[B]^y$$

x = the order of the reaction with respect to reactant A

y = the order of the reaction with respect to reactant B

k = the rate constant

On the GSE, you may be required to solve for the rate constant. Let's look at the following example:

$$BrO_3^- + 5Br^- + 6H^+ \rightarrow 3Br_2 + 3H_2O$$

You might be provided with a table of experimental results that looks like the following:

Experiment #	[BrO₃⁻]	[Br⁻]	[H⁺]	Initial rate (mol/sec)
1	0.10	0.10	0.10	8.0×10^{-4}
2	0.20	0.10	0.10	1.6×10^{-3}
3	0.20	0.20	0.10	3.2×10^{-3}
4	0.10	0.10	0.20	3.2×10^{-3}

In order to determine the *order* of each reactant, we need to look at the effect of each reactant on the reaction.

Let's look at the first reactant, BrO_3^-. We can tell from experiments 1 and 2 that the concentration of BrO_3^- has doubled, while the concentrations of the other two reactants have not changed. The overall rate of the reaction has also doubled from experiment 1 to 2. Therefore, we say that the reaction rate changed linearly with the concentration of BrO_3^-, and that the reaction is first order in BrO_3^-, written $[BrO_3^-]^1$.

Next, we notice that between experiments 2 and 3, the concentration of $[Br^-]$ has doubled while the concentrations of the other reactants are unchanged. The reaction rate from experiment 2 to experiment 3 has also doubled, which means that the rate has changed linearly with the concentration of Br^-. The reaction is first order with respect to Br^-, written $[Br^-]^1$.

Finally, we look at the last reactant, H^+. We notice that between experiments 1 and 4, the concentration of H^+ has doubled while the concentrations of the other two reactants have not changed. The reaction rate between experiments 1 and 4 has quadrupled, which means that the rate changes with $[H^+]$ are squared. The reaction is said to be second order with respect to H^+, written $[H^+]^2$.

In order to calculate, k we can manipulate the rate law equation.

$$\text{Rate} = k\,[BrO_3^-][Br^-][H^+]^2$$

$$k = \frac{\text{Rate}}{\left[BrO_3^-\right]\left[Br^-\right]\left[H^+\right]^2}$$

Now we can use the results from any of the experiments to figure out k. Let's use the results from experiment 1.

$$k = \frac{\text{Rate}}{\left[BrO_3^-\right]\left[Br^-\right]\left[H^+\right]^2}$$

$$k = \frac{8.0 \times 10^4}{[0.10][0.10][0.10]^2}$$

$$k = 8.0$$

The units of k are $L^3/mol^3 \cdot s$.

EQUILIBRIUM

As you're probably already aware, reactions run not only in the forward direction, in which reactants form a product, but they can also run in reverse, where the products react to reform reactants. Reactions that run in both directions are said to be **reversible** and usually occur in closed systems. When a reaction begins, the forward rate is greater than the reverse rate. As the amount of products increases, the rate of the reverse reaction starts to catch up to the rate of the forward reaction. A reaction reaches **chemical equilibrium** when reactants and products are formed at the same rate so that their concentrations do not change.

In order to calculate the concentrations of the reactants and products at equilibrium, we need to know the reaction's **equilibrium constant (K_{eq})**.

The equilibrium constant is different for every reaction, and you will not be required to memorize any equilibrium constants for the GSE. Sometimes K will appear in different forms:

K_a = equilibrium constant for acids

K_b = equilibrium constant for bases

K_i = equilibrium constant for ionization

K_{sp} = solubility product

Take a look at the following general reaction:

$$dD + eE \leftrightarrow fF + gG$$

$$\frac{[F]^f [G]^g}{[D]^d [E]^e}$$

To set up an equilibrium equation, follow the rules listed below:

- Products are placed in the numerator and are multiplied by each other.

- Reactants are placed in the denominator and are multiplied by each other.

- The coefficient of each substance becomes its exponent.

- Pure solids and liquids are not included in the equation, but aqueous molecules *are* included.

- If the reactants and products are gases, then partial pressures are used instead of molar concentrations.

Let's look at the following equation:

$$N_2 + O_2 \leftrightarrow 2NO$$

The equilibrium constant for the equation above is represented by the following expression:

$$K_{eq} = \frac{[NO]^2}{[N_2][O_2]}$$

The value of the K_{eq} tells us which direction is favored in the reaction. If K_{eq} is larger than one, then the forward direction is favored and products are present in larger amounts. If K_{eq} is less than one, then the reverse direction is favored and reactants are present in larger amounts.

LE CHATELIER'S PRINCIPLE

If something disturbs a reaction in equilibrium, such as the addition or removal of reactant or product, the reaction will adjust itself in the direction that will offset the disturbance. This concept is known as **Le Chatelier's principle**.

Take a look at the following equation:

$$CO(g) + 3H_2(g) \leftrightarrow CH_4(g) + H_2O(g)$$

What will happen if we remove the water vapor produced by this reaction? Well, we will create more room for the production of water vapor, and it is predicted by Le Chatelier's principle that the forward reaction will be temporarily favored. Thus the equilibrium will shift to the right.

What happens if we *add* water vapor to this reaction? Well, this means that the right side of the equation now contains substrate in excess. Le Chatelier's principle tells us that the equilibrium will shift to the left, and the reverse reaction will be temporarily favored.

Similar effects will take place if reactants are added or removed. Keep in mind that since neither solids nor liquids are included in the equilibrium calculation, the addition or removal of a pure solid or liquid will have no effect on the state of equilibrium.

If the volume in which the reaction is taking place is decreased, the reaction will shift in the direction that yields fewer moles of gas in order to accommodate the smaller volume. In the reaction above, the forward direction will be favored if volume is decreased since there are fewer moles of gas produced in the forward direction than in the reverse. Remember from the ideal gas equation ($PV = nRT$), that volume and pressure are inversely related. Therefore, if you're told that pressure is increased, you need to immediately conclude that volume is decreased and that the reaction will shift toward the side that produces the fewer number of moles. If volume is increased or pressure is decreased, the reaction will favor the direction that yields more moles of gas in order to fill the increased capacity. In our reaction above, the reverse reaction will be favored if volume is increased, since it produces more moles of gas.

A change in volume will not have an effect on equilibrium when the reaction does not contain gases as reactants or products, or the same number of moles are produced on both sides of the equation.

Equilibrium equations are also affected by temperature. If the temperature of a reaction increases, the reaction will shift in the direction in which heat is absorbed. If the temperature decreases, the reaction will shift in the direction in which heat is released.

SOLUBILITY PRODUCT

As we mentioned earlier in our review, sodium chloride (table salt) dissolves easily in water. However, some salts do not dissolve as readily as NaCl, and these salts are considered slightly soluble or insoluble—they do not completely dissociate in solution. The solubility product (K_{sp}) expresses the extent to which a salt will dissociate in water. The greater the solubility product, the greater the solubility of the salt.

The following equation represents the dissociation of barium oxide in water.

$$Ba(OH)_2 \;(s) \rightarrow Ba^{2+} \;(aq) + 2\;OH^- \;(aq)$$

The equilibrium equation is written as follows:

$$K_{sp} = [Ba^{2+}][OH^-]^2$$

$$K_{sp} = 5.0 \times 10^{-3}$$

Barium oxide is left out of the equation because it's a solid—it does not effect equilibrium.

If you're given the solubility product for a solid, you can calculate the relative amount of the solid that will dissolve in solution. Let's look at that barium oxide equation again:

$$Ba(OH)_2 \;(s) \rightarrow Ba^{2+} \;(aq) + 2\;OH^- \;(aq)$$
$$x \qquad\qquad \rightarrow x \qquad\quad 2x$$

We know that the solubility product is equal to:

$$K_{sp} = [Ba^{2+}][OH^-]^2$$

$$5.0 \times 10^{-3} = (x)(2x)^2 = 4x^3$$

Since you're not allowed to use a calculator on the GSE, you'll need to approximate the value of x. You could either use the answer choices and backsolve, or you could estimate that x is slightly greater than 0.1, which means the that a solution of $Ba(OH)_2$ can never be more concentrated than approximately 0.1 M.

90 ◆ CRACKING THE GOLDEN STATE EXAMINATION: CHEMISTRY

COMMON ION EFFECT

To illustrate the common ion effect, let's take another look at the solubility expression for barium hydroxide, $Ba(OH)_2$.

$$Ba(OH)_2 \ (s) \rightarrow Ba^{2+} \ (aq) + 2 \ OH^- \ (aq)$$

What happens if we add sodium hydroxide (NaOH) to the solution? Well, we'll have excess hydroxide ions (OH^-) floating around, causing a strain on the equilibrium. Remember Le Chatelier's principle? In order to adjust for the excess OH^- ions, the equilibrium will shift to the left, causing $Ba(OH)_2$ to precipitate out of solution. This phenomenon is known as the **common ion effect**. It does not matter from what substance the excess OH^- ions came, they will affect the equilibrium.

REACTION QUOTIENT

The **reaction quotient (Q)** expression has the same format as the equilibrium constant equation, but its value does not necessarily represent the equilibrium state. Instead, the reaction quotient is used as a tool to determine the direction in which the equilibrium will shift under certain conditions. The reaction quotient represents potential concentrations.

Let's look again at the barium hydroxide equilibrium constant equation.

$$Ba(OH)_2 \ (s) \rightarrow Ba^{2+} \ (aq) + 2 \ OH^- \ (aq)$$
$$x \quad\quad \rightarrow x \quad\quad + 2x$$
$$0.5 \ M \quad \rightarrow 0.5 \ M \quad + 2(0.5 \ M)$$

Would it be feasible for us to make a $0.5 \ M$ solution of $Ba(OH)_2$? A $0.5 \ M$ solution of $Ba(OH)_2$ would contain $0.5 \ M$ of Ba^{2+} ions and 2(0.5 moles) of OH^- ions. Therefore, the reaction quotient is equal to:

$$Q = [Ba^{2+}][OH^-]^2 = (0.5)(2 \times 0.5)^2 = 5.0 \times 10^{-1}$$

Since Q is greater than the K_{sp} (5×10^{-3}) not all of the barium hydroxide will dissolve.

Here are the rules for determining the solubility of solutes using the reaction quotient (Q):

- If Q is less than the K_{sp} for a reaction, then the reaction will continue in the forward direction and all of the product will be in solution.

- If Q is greater than the K_{sp} for a reaction, then the reaction will shift in a direction to form reactants and cause a precipitate.

- If Q is equal to K_{sp}, then the reaction will be at equilibrium.

KINETICS AND EQUILIBRIUM

1. Three experiments were performed and the following results were obtained regarding the product formation rates.

$$A + B \rightarrow C$$

Experiment	[A]	[B]	Reaction rate (mol/L-sec)
1	0.2	0.4	3.0×10^{-3}
2	0.2	0.8	6.0×10^{-3}
3	0.4	0.4	1.2×10^{-2}

What is the rate law for the reaction?

A. Rate = $k[A]^2$

B. Rate = $k[A][B]$

C. Rate = $k[A]^2[B]$

D. Rate = $k[A][B]^2$

Answer:

C. Comparing experiments 1 and 2, when the concentration of [A] is held constant and the concentration of [B] is doubled the rate of reaction is also doubled. The reaction is, therefore, first order with respect to [B]. Comparing experiments 1 and 3, when the concentration of [B] is held constant and the concentration of [A] is doubled the reaction rate quadruples. The reaction is second order with respect to [A]. The rate law is then = $k[A]^2[B]$.

2. $CO(g) + 2H_2(g) \leftrightarrow CH_3OH(g)$

Which of the following changes to the reaction above would result in the decrease of the concentration of CH_3OH?

A. The addition of CO to the system

B. The addition of H_2 to the system

C. The volume of the system is increased at constant temperature

D. The volume of the system is decreased at constant temperature

Answer:

C. According to LeChatelier's principle, the equilibrium of a reaction will shift in the direction that will reduce the stress that has been placed on the system. If volume is increased the equilibrium will shift towards the side that has fewer moles of gas. In this system, the side with fewer moles of gas is the left side. A shift to the left will cause a decrease in the amount of CH_3OH.

Adding reactants (A) and (B) will cause the equilibrium to shift to the right, thus increasing the concentration of CH_3OH.

3. If $1.0\ mol \times 10^{-3}$ of lead (II) iodide, PbI_2, dissolves in 1L of water at 25 °C, what is solubility product (K_{sp})?

A. 1×10^{-9}

B. 2×10^{-9}

C. 1×10^{-6}

D. 2×10^{-6}

Answer:

B. The reaction is as follows:

$PbI_2(s) \quad Pb^{2+}(aq) + \quad 2I^-(aq)$

$1 \times 10^{-3} \qquad 1 \times 10^{-3} \qquad\qquad 2[1 \times 10^{-3}]$

$K_{sp} = [Pb^+][I^-]^2 = [1 \times 10^{-3}][2(1 \times 10^{-3})]^2 = 4 \times 10^{-9}$

4. $CaC_2O_4(s) \leftrightarrow Ca^{2+}(aq) + C_2O_4^{2-}(aq)$

Calcium oxalate, CaC_2O_4, dissolves in water to produce Ca^{2+} and $C_2O_4^{2-}$ ions in solution. Which of the following changes to the system, will cause calcium oxalate to precipitate out of solution?

A. Addition of calcium chloride, $CaCl_2$

B. Addition of water

C. Removal of $C_2O_4^-$ ions

D. Removal of Ca^{2+} ions

Answer:

A. Adding calcium chloride will introduce additional Ca $^{2+}$ ions into the solution, thus causing a stress on the system. To relieve the stress caused by the additional Ca $^{2+}$ ions the reaction will shift to the left, causing calcium oxalate to precipitate out of solution.

The removal of the products (C) and (D) will cause the reaction to shift to the right.

5. $CaCO_3(s) \leftrightarrow Ca^{2+}(aq) + CO_3^{2-}$

Calcium carbonate dissolves in aqueous solution to yield Ca^{2+} and CO_3^{2-} ions. If the solubility product (K_{sp}) is equal to 2.8×10^{-9} what is the concentration of Ca^{2+} ions in a saturated 1.0 liter solution of $CaCO_3$ at room temperature?

A. $2.8 \times 10^{-9} M$
B. $5.3 \times 10^{-9} M$
C. $2.8 \times 10^{-5} M$
D. $5.3 \times 10^{-5} M$

Answer:

D. $CaCO_3(s) \leftrightarrow Ca^{2+}(aq) + CO_3^{2-}$
   $\quad x \qquad\qquad x \qquad\quad x$

$$K_{sp} = [Ca^{2+}][CO_3^{2-}] = [x][x] = x^2$$

$$x = [Ca^{2+}] = \sqrt{K_{sp}}$$

$$x = \sqrt{2.8 \times 10^{-9}}$$

Approximate:

$$\sqrt{2.8 \times 10^{-9}} \approx \sqrt{25 \times 10^{-10}} = 5 \times 10^{-5}$$

Chapter 10

ACIDS AND BASES

Acids and bases were originally detected through the sense of taste. A sour substance is indicative of acidic content, while bitter substances are often basic. Svante Arrhenius, a Swedish chemist, was the first to define the terms acid and base. The Arrhenius definition of an acid is *a substance that reacts with water to produce the hydronium ion (H_3O^+)*. Look at the following reaction:

$$HCl + H_2O \rightarrow H_3O^+ + Cl^-$$

In this reaction, hydrochloric acid reacts with water to yield the hydronium ion and a chlorine ion. Therefore, it is an acid.

Also according to Arrhenius, a base *is a substance that produces hydroxide ions (OH^-) in water, or can react with hydronium ions*. Take a look at the following reaction:

$$NH_3 + H_2O \rightarrow NH_4^+ + OH^-$$

Ammonia is ionized into an ammonium ion and a hydroxide ion; it meets the criteria for an Arrhenius bases.

Johannes Brönsted and **Thomas Lowry** later came up with a different way of defining acids and bases—you should be familiar with both the Arrhenius and

95

the Brönsted-Lowry definitions for the Chemistry GSE. Brönsted and Lowry defined an acid as *a substance that is a proton donor*. Let's look again at the hydrochloric acid reaction:

$$HCl \rightarrow H^+ + Cl^-$$

The hydrogen ion has one proton in its nucleus, as is usual for hydrogen, but no electrons. Therefore, hydrochloric acid has donated a proton; it's a Brönsted-Lowry acid.

Let's look at the ammonia reaction again:

$$NH_3^+ + H_2O \rightarrow NH_4^+ + OH^-$$

In this reaction, ammonia picks up an additional proton to become an ammonium ion; it's a base according to the Brönsted-Lowry definition—*a substance that can accept a proton in solution*.

The final name you should be familiar with is **G. N. Lewis**. Lewis defined acids and bases according to the movement of electron pairs in reactions, basically saying that *an acid is an electron pair acceptor, and a base is an electron pair donor*. Lets look one more time at the ammonia reaction:

This time we wrote the reaction with the Lewis dot structures. As you can see, ammonia starts off with a pair of electrons, but when it dissociates in water, it donates its electron pair and bonds with a hydrogen atom. According to Lewis, since ammonia donates its electron pair to a bond, it is a base.

CONJUGATE ACID-BASE PAIRS

In an acid–base reaction, the two substances (meaning ions or molecules) whose formulas differ by only one H^+ are known as a **conjugate acid–base pair**.

As we said, a Brönsted-Lowry acid will dissociate into a hydrogen ion and its conjugate base. Lets look at the following reaction, involving acetic acid:

$$HC_2H_3O_2 + H_2O \rightarrow C_2H_3O_2^- + H^+$$

Acetic acid, $HC_2H_3O_2$, dissociates into a hydrogen ion and its conjugate base acetate ($C_2H_3O_2^-$). Acetic acid and acetate are a conjugate acid-base pair.

This will probably come as no surprise to you, but a Brönsted-Lowry acid has one more H⁺ than its conjugate base does, and a Brönsted-Lowry base has one less hydrogen than its conjugate acid. Some examples of conjugate pairs are shown in the chart below.

ACID	BASE
$HClO_4$	ClO_4^-
HF	F^-
NH_4^+	NH_3
H_3O^+	H_2O
H_2O	OH^-

The stronger the acid, the weaker the conjugate base. The stronger the base, the weaker the conjugate acid.

WEAK ACIDS

Recall that the equilibrium constant (K_{eq}) is a measure of product and reactant concentrations when a reaction has reached equilibrium.

$$aA + bB \leftrightarrow cC + dD$$

$$K_{eq} = \frac{[C]^c[D]^d}{[A]^a[B]^b}$$

When dealing with acids and bases, we look at the dissociation constants (K_a and K_b) which reveal the extent to which acids dissociate in solution (a measure of their strength), as well as the extent to which bases accept hydrogen ions in solution.

Let's talk about acids first. The value of the acid dissociation constant (K_a) reveals the strength of the acid. The greater the K_a, the stronger the acid and the more hydrogen ions that dissociate into solution. Look at the following acid dissociation constants:

Boric acid $\quad K_a = \dfrac{[H^+][H_2BO_3]}{[H_3BO_3]} = 5.9 \times 10^{-10}$

Phosphoric acid $\quad K_a = \dfrac{[H^+][HPHO_3]}{[H_2PHO_3]} = 1.6 \times 10^{-2}$

Phosphoric acid has a greater K_a value, which means that it dissociates a greater number of hydrogen ions than the boric acid in solution. Although phosphoric acid is the stronger of the two acids, it is not considered a particularly strong acid. Strong acids such as HCl do not have measurable dissociation constants. Strong acids completely dissociate in solution; this means that they never reach equilibrium.

Take a look at the following equation, showing the complete dissociation of HCl:

$$HCl \rightarrow H^+ + Cl^-$$

$$K_{eq} = \frac{[H^+][Cl^-]}{[HCl]}$$

Since HCl dissociates completely, its concentration should be infinitesimally small. This means that K_{eq} would go to infinity. Also, since HCl is a strong acid, its conjugate base (Cl^-) must be a weak base. You can figure this out—the Cl^- ion does not have the strength to accept protons and drive this reaction in the reverse direction.

Other examples of strong acids are HBr, HI, HNO_3, $HClO_4$, and H_2SO_4.

The idea is the same for strong bases. Strong bases dissociate completely, producing very weak conjugate acids. Examples of strong bases are hydroxides of alkali metals and alkaline earth metals (except Be) such as LiOH, NaOH, KOH, $Ca(OH)_2$, and $Ba(OH)_2$.

Some acids are able to donate more than one hydrogen ion in solution. These substances are known as **polyprotic acids**. Some examples of polyprotic acids are carbonic acid (H_2CO_3), oxalic acid ($H_2C_2O_4$), phosphoric acid (H_3PO_4), and sulfuric acid (H_2SO_4). A polyprotic acid is more eager to donate its first hydrogen ion than subsequent ions. This is reflected in the diminishing K_a values. Take a look at the dissociations of phosphoric acid:

$$H_3PO_4 \rightarrow H^+ + H_2PO_4^- \qquad K_a = 6.9 \times 10^{-3}$$
$$H_2PO_4^- \rightarrow H^+ + HPO_4^{2-} \qquad K_a = 6.2 \times 10^{-8}$$
$$H_2PO_4^{2-} \rightarrow H^+ + PO_4^{3-} \qquad K_a = 4.8 \times 10^{-13}$$

Notice in the equations above, both $H_2PO_4^{2-}$ and HPO_4^{2-} can either accept a proton or donate a proton. Substances that can act as either an acid or a base are known as **amphoteric** substances. Polyprotic acids yield conjugate bases that are amphoteric.

PH

The concentration of an acid or a base in solution can be expressed by **pH** and **pOH** values. The pH is a measure of the hydrogen ion concentration in a solution, and pOH is a measure of the hydroxide ion concentration in solution.

$$pH = -\log[H^+]$$

$$pOH = -\log[OH^-]$$

These measures allow us to represent acid or base character on a scale of 0 to 14. A solution is acidic when it has a pH value less than 7—in other words, the $[H^+]$ is greater than the $[OH^-]$ in solution. A solution is basic when it has a pH that's greater than 7—the $[H^+]$ is less than the $[OH^-]$ in solution.

As $[H^+]$ increases, the pH of a solution decreases. In the same way, as $[H^+]$ decreases, pH increases.

The pH of strong acids is equal to their concentration, since these acids dissociate completely in solution. For example, a 0.001M solution of HCl dissociates completely into H^+ and Cl^- ions.

HCl	$\rightarrow$	H^+	+	Cl^-
0.001M		0.001M		0.001M

In this case, the pH is calculated directly from the molarity of the acid solution.

$$pH = -\log[H^+]$$

$$[H^+] = 0.001 \text{ M} = 1.0 \times 10^{-3} \text{ M}$$

$$pH = 3$$

In the case of weaker acids, equilibrium occurs between the acid and the ions into which it dissociates. Since weak acids are not as eager to contribute hydrogen ions to solution, the concentration of the acid itself will be greater than the concentration of $[H^+]$. Let's try an example that involves the calculation of pH of a weak acid. Let's say that we wanted to find the pH of a 0.4 M solution of hydrofluoric acid (HF). (The K_a for HF is 6.8×10^{-4}).

Lets assume that x moles of HF dissociate into x moles of H^+ and x moles of F^-.

	HF	$\rightarrow$	H^+	+	F^-
initial:	0.4 M		0 M		0 M
equilibrium:	$(0.4 - x)$ M		x M		x M

We set up the equilibrium as follows:

$$K_a = \frac{[H^+][F^-]}{[HF]}$$

$$6.8 \times 10^{-4} = \frac{(x)(x)}{0.4 - x}$$

Now, we can assume that not much of the HF will dissociate, since its K_a is so small. This means that the value of x will be quite a bit smaller than $0.4\,M$, and the value of $0.4 - x$ will be very close to $0.4\,M$. We can then replace $0.4 - x$ with 0.4, and we're left with the following equation:

$$\frac{x^2}{0.4} = 6.8 \times 10^{-4}$$

From this equation we can perform rough estimations.

$$x^2 = (6.8 \times 0.4) \times 10^{-4}$$

$$x^2 \approx (7 \times 0.4) \times 10^{-4}$$

$$x^2 \approx 2.8 \times 10^{-4}$$

$$x \approx 1.6 \times 10^{-2}$$

The hydrogen concentration will be a little greater than 1.0×10^{-2}. Taking the negative log of this value, we see that the pH is slightly less than 2.

ACID AND BASE SALTS

As you know, when a salt dissolves in solution, it dissociates into its component ions. The resulting salt solution can be neutral, acidic, or basic, depending upon the identities of the ions.

A salt solution is *neutral* when the dissociated ions are conjugates of a strong acid and a strong base. For example, sodium chloride (NaCl) dissociates into Na^+ and Cl^- ions. Na^+ is the conjugate acid of a strong base, NaOH, and Cl^- is the conjugate base of a strong acid, HCl. The resulting salt solution is neutral, and has a pH of 7.

$$NaCl \rightarrow Na^+ + Cl^- \qquad pH = 7$$

A salt solution is *acidic* when the dissociated ions are conjugates of a strong acid and a weak base. For example, ammonium chloride, NH_4Cl, dissociates into NH_4^+ and Cl^- ions. NH_4^+ is a conjugate acid from a weak base, NH_3, and as we just saw, Cl^- is a conjugate base from a strong acid, HCl. The resulting salt solution is acidic; its pH is less than 7. Ammonium chloride is known as an **acid salt**.

$$NH_4Cl^- \rightarrow NH_4^+ + Cl^- \quad pH < 7$$

A salt solution is *basic* when the dissociated ions are conjugates of a weak acid and a strong base. For example, sodium acetate ($NaC_2H_3O_2$) dissociates into Na^+ and $C_2H_3O_2^-$ ions. Na^+ is a conjugate acid of a strong base, and $C_2H_3O_2^-$ is a conjugate base of a weak acid ($HC_2H_3O_2$) The resulting salt solution is basic; its pH is greater than 7. Sodium acetate is known as a **base salt**.

$$NaC_2H_3O_2 \rightarrow Na^+ + C_2H_3O_2^- \quad pH > 7$$

NEUTRALIZATION

After eating a big bowl of spicy chili, some of us may experience the phenomenon of heartburn, a result of acid reflux from the stomach up into the esophagus. In order to neutralize the stomach acid, you might take an antacid. These antacids contain bases that will react to neutralize the acid.

But what is the product of the reaction between an acid and a base? Well, the answer is a salt and water. Look at the following reaction:

$$HCl + NaOH \leftrightarrow NaCl + H_2O$$

The conjugate strong acid and base of HCl and NaOH, respectively, bond to form a neutral salt. Keep this in mind for the exam: *When an acid and base react, the products are a salt and water.*

TITRATIONS

It is possible to determine the concentration of an acid or base in a solution by performing a **titration**. For instance, in order to determine the concentration of an acid in solution, the solution can be titrated with a basic solution of known concentration. If we had a 1-L solution of HCl of unknown concentration, we would start the titration by adding 0.1 *M* NaOH to the solution in small quanities with a buret. An indicator, such as phenolphthalein, would tell us when we have reached the equivalence point in the titration. Phenolphthalein is clear at a pH below 7, and pink at a pH above 7. The **equivalence point** is the point at which enough NaOH has been added to completely neutralize the HCl. If 0.6 L of 0.1 *M* NaOH was required in order to reach the equivalence point, then the original concentration of HCl could be calculated as follows:

$$\text{Molarity} = \frac{\text{moles}}{\text{liter}}$$

$$\text{Moles} = (\text{molarity})(\text{liter})$$

$$\text{Moles of NaOH added} = (0.1\,M)(0.6\,L) = 0.06$$

Since there is a 1:1 molar ratio in HCl and NaOH, there must have been 0.06 moles of HCl in solution. Therefore, the concentration of HCl must have been:

$$\text{Molarity of the HCl solution} = \frac{\text{moles}}{\text{liter}} = \frac{0.06\,\text{mol}}{1\,L} = 0.06\,M$$

The following diagram represents a titration curve for an acid titration like the one we just looked at. A titration curve is a plot of the pH versus the amount of titrant (which can be either acid or base). The equivalence point in the reaction is labeled.

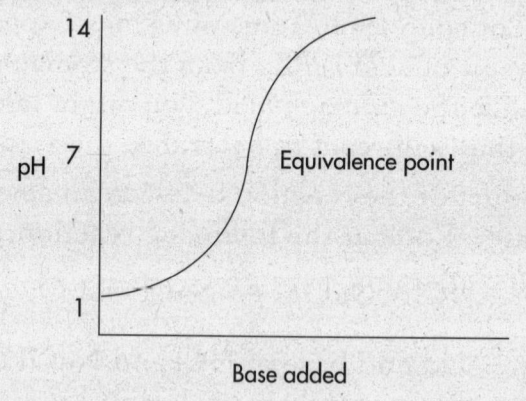

If we were titrating a basic solution of unknown concentration with an acid solution of known concentration, we would end up with the following titration curve:

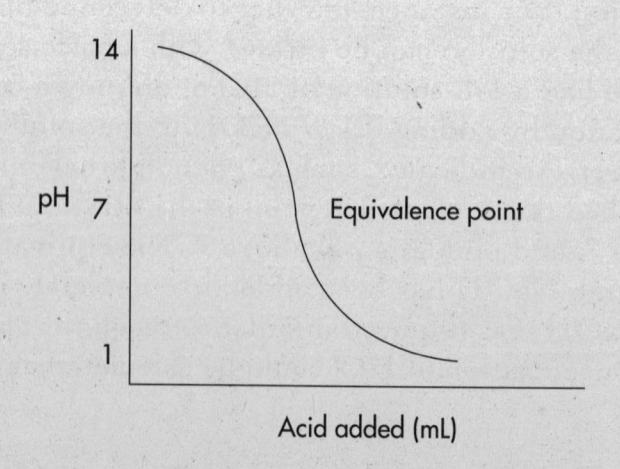

If we were neutralizing a polyprotic acid, such as carbonic acid (H_2CO_3), the titration curve would show two equivalence points, one for the neutralization of each of the dissociating hydrogens of carbonic acid.

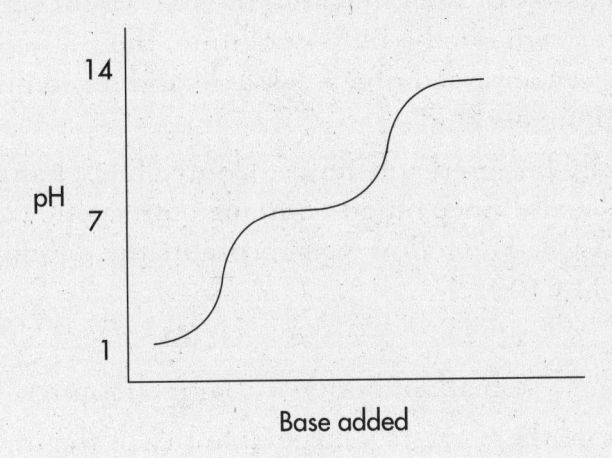

For the same reason, if we were neutralizing phosphoric acid (H_3PO_4), the titration curve would show three equivalence points.

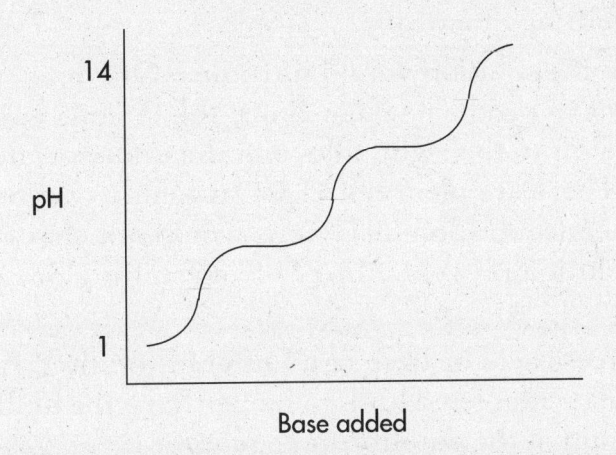

Earlier, we discussed that an acid and a base react to form water and a salt. The equivalence point of a titration represents the time at which the pH of the solution is equal to the pH of the conjugate salt. In our titration example involving HCl and NaOH, the salt formed by the conjugates was NaCl, a neutral salt. Therefore, the equivalence point would occur when the pH equals 7. If the salt formed is acidic, then the equivalence point occurs when the pH is less than 7. If the salt formed is basic, the equivalence point would occur when the pH is greater than 7.

BUFFERS

A **buffer** is the component of a solution that renders the solution capable of resisting changes in pH when small amounts of acid or a base are added to it. If we added a small quantity of HCl to water, the pH would significantly decrease. However, if the water contained a buffer solution, the pH would not change as drastically. A buffer is comprised of a weak acid and its conjugate base, or a weak base and its conjugate acid.

Blood is a naturally buffered substance. Controlling changes in the pH of blood is crucial, especially since blood contains both acetic acid (H_2CO_3) and its conjugate base (HCO_3^-). The following represents the equilibrium equation of the buffer solution (H_2CO_3):

$H_2CO_3 \leftrightarrow$ $H^+ +$ HCO_3^-
(large amount)   (small amount)   (large amount)

In a buffer solution, there exists a large quantity of the undissociated acid and its conjugate.

Adding an acid to the solution would introduce additional H^+ ions. These additional H^+ ions would combine with the HCO_3^- ions and shift the equilibrium to the left. There is enough HCO_3^- to consume the additional H^+ ions and prevent the pH from shifting drastically.

Adding a base to the solution would introduce OH^- ions, which would combine with the H^+ ions to produce water. Since the H^+ ions would be removed, the equilibrium would shift to the right, allowing the undissociated acid to dissociate into more H^+ ions. There are plenty of H_2CO_3 atoms to dissociate, which would mean that the total concentration of H^+ ions would not change significantly. This is how the buffer solution prevents the pH from shifting drastically with the addition of a base.

When adding excess acid or base to a buffered solution, there is a point at which the buffer will be depleted and no longer have the ability to resist changes in pH. Let's look again at the equilibrium equation:

H_2CO_3 $\leftrightarrow$ $H^+ +$ HCO_3^-
(large amount)   (small amount)   (large amount)

$$K_a \text{ of } H_2CO_3 = \frac{[H^+][HCO_3^-]}{[H_2CO_3]} = 4.3 \times 10^7$$

Since we have added equal concentrations of the undissociated acid (H_2CO_3) and its conjugate base (HCO_3^-) their concentrations will cancel each other out. This means that we are left with the following equation:

$$K_a = [H^+]$$

Therefore:
$$pK_a = pH$$

For an ideal buffer, the pH will equal the pK_a of the acid from which it was formed. When creating a buffer, it is best to select an acid whose pK_a value is close to the pH of the solution you're targeting.

ACIDS AND BASES

1. Which of the following is NOT an amphoteric substance?

    A. HSO_4^-

    B. $HC_2O_4^-$

    C. H_2O

    D. HNO_3

    Answer:

    D. Be careful…this question asks which is NOT an amphoteric substance. An amphoteric substance can act as either a base or an acid. When it acts as a base it will gain an H^+ and when it acts as an acid it will lose an H^+.

    HSO_4^- can lose an H^+ and become SO_4^{2-} or it can gain an H^+ to become H_2SO_4

    $HC_2O_4^-$ can lose an H^+ to become $C_2O_4^{2-}$ or it can gain an H^+ to become $H_2C_2O_4$

    H_2O can lose an H^+ to become OH^- or it can gain an H^+ to become H_3O^+

    HNO_3 can only act as an acid and lose an H^+ to become NO_3^-. Therefore, D is the answer.

2. Which of the following is a polyprotic acid?

    A. H_2SO_4

    B. HNO_3

    C. $HClO_4$

    D. $HC_2H_3O_2$

    Answer:

    A. A polyprotic acid can give up more than one hydrogen ion. Although H_2SO_4 is more willing to give up its first H^+ to become HSO_4^-, it can also give up another H^+ to become SO_4^{2-}.

3. If a laboratory technician were looking to neutralize a 100 mL sample of 0.1M NaOH, how much 0.5M HCl would she need to add?

A. 5 mL

B. 20 mL

C. 100 mL

D. 120 mL

Answer:

B. First start by using process of elimination. If we need to neutralize 100 mL of a 0.1M NaOH solution with a solution of HCl at a greater concentration, it would require less HCl than the original 100 mL of NaOH. Therefore, cross off any answer choice that is not less than 100 mL. You can cross off C and D.

The moles of HCl and NaOH will be equal once the solution is neutralized. So first calculate what we know, the moles of NaOH.

$$\text{Molarity} = \text{moles/volume}$$

$$\text{moles} = \text{Molarity} \times \text{volume} = (0.1M)(100 \text{ mL}) = 10 \text{ mmol}$$

Next calculate the volume of HCl that will need to be added to obtain 10 mmol of HCl.

$$\text{volume} = \text{moles/Molarity} = 10 \text{ mmoles}/0.5 \text{ M} = 20 \text{ mL}$$

4. The acid dissociation constant (K_a) for acetic acid ($HC_2H_3O_2$) is 1.7×10^{-5}. What is the value of the base dissociation constant (K_b) for $C_2H_3O_2^-$?

A. 1.7×10^{-9}

B. 5.88×10^{-9}

C. 1.7×10^{-8}

D. 5.88×10^{-8}

Answer:

D. $(K_a)(K_b) = 1 \times 10^{-14}$ for conjugate pairs. Therefore,

$$K_b = \frac{1 \times 10^{14}}{K_a} = \frac{1 \times 10^{14}}{1.7 \times 10^{-5}}$$

Use approximations to do the math:

$$K_b = \frac{10 \times 10^{-15}}{2 \times 10^{-5}} = 5 \times 10^{-10}$$

5. What is the acid dissociation constant (K_a) of a 1-molar acid solution with a pH of 5?

$$K_a = \frac{[H^+][A^-]}{[HA]} = \frac{[1 \times 10^{-5}][1 \times 10^{-5}]}{1} = 1 \times 10^{-10}$$

A. 1×10^{-11}

B. 1×10^{-10}

C. 1×10^{-5}

D. 1×10^{-1}

Answer:

B. If an acid solution has a pH of 5, the concentration of H^+ is 1×10^{-5}

$$pH = -\log[H^+]$$

$$[H^+] = 1 \times 10^{-5}$$

The following equation represents the weak acid at equilibrium:

$$HA \leftrightarrow H^+ + A^-$$

For every HA that dissociates there is one H^+ and one A^-; therefore, $[H^+] = [A^-]$

For weak acids, we can assume that not much HA dissociates and that the concentration of HA stays very close to 1 molar.

Chapter 11

THERMODYNAMICS

Thermodynamics is the study of energy and entropy changes that take place during physical and chemical events. According to the *first law of thermodynamics*, the total amount of energy in the universe remains constant. In other words, energy cannot be created or destroyed. Energy is exchanged in a system in two forms: as heat or work. **Heat** is a form of energy that moves from a system of higher temperature to a system of lower temperature, and **work** is energy that results in the displacement of an object ($W = f \times d$).

ENTROPY

According to the *second law of thermodynamics*, the total entropy of the universe will increase when a process is spontaneous. **Entropy** (S) is the measure of randomness or disorder of a system. Let's consider the process of an ice cube melting. Prior to melting, the ice cube is in an ordered crystalline structure in which the water molecules are in fixed positions. The entropy of the ice cube is relatively low. When the ice melts, the water molecules are able to move around; their structure is less ordered. The entropy of liquid water is greater than that of ice. Changes in entropy (ΔS) can be calculated by subtracting the initial entropy of the reactants (S_r) from the final entropy of the products (S_p).

$$\Delta S = S_p - S_r$$

The ΔS is positive if the products have higher entropy than the reactants. If the products have lower entropy than the reactants, then the ΔS is negative. Generally speaking, liquids have more entropy than solids, and gases have more entropy than liquids.

Let's consider the melting ice cube again. The ΔS for ice melting will be positive, since the product is of higher entropy than the reactant. If we were to freeze the ice, the ΔS would be negative since we are going from a less ordered state to a more ordered state. The two values of ΔS will be the same, but they will have opposite signs.

ENTHALPY

Enthalpy (H) is the heat content of a closed system. Enthalpy changes when substances react and bonds are formed or broken. Changes in enthalpy (ΔH) are calculated in the same way as changes in entropy. The enthalpy of the reactants is subtracted from the enthalpy of the products. This is also known as the **heat of the reaction**.

$$\Delta H = H_p - H_r$$

If the products have less enthalpy than the reactants, then ΔH is negative and heat is released in the course of the reaction; the reaction is **exothermic**. If the products have more enthalpy than the reactants, then ΔH is positive and heat is absorbed in the course of the reaction; the reaction is **endothermic**.

Substances prefer lower energy states, so reactions are more likely to occur when the products have less energy ($-\Delta H$) than the reactants. When ΔH is positive, energy is required to break the bonds of a stable molecule. The value of H will be the same for both the formation of bonds ($-\Delta H$) and for the breaking of bonds ($+\Delta H$).

MAKING AND BREAKING BONDS

The total amount of energy released when a bond is formed is known as the **bond energy** (ΔH_f°) of a product. The heat of formation of pure elements in their uncombined, stable form is zero. For example, the heat of formation of C(s) is zero. The stable, pure form of oxygen is diatomic oxygen, O_2. Therefore, the ΔH_f° of O_2 equals zero, while the ΔH_f° of O(g) is 247.5 kJ/mol.

The ΔH_f° is negative when energy is *released* in the course of a "formation reaction" in which a compound is formed from pure elements. The ΔH_f° is positive when energy is *consumed* in the course of a "formation reaction" in which a compound is formed from pure elements.

The amount of energy required to break a bond is known as the **bond dissociation energy**. As we mentioned earlier, the amount of energy released during the formation of a bond is equal to the amount of energy that will be required to break the bond. Therefore, the bond dissociation energy is the same as the bond energy.

It is important to remember that ΔH is a constant value. When you are determining the total amount of heat that's released or absorbed in a reaction, the ΔH can be thought of as being as constant as any of the other products. This means that if the reactants in an *exothermic* reaction are doubled, then twice the amount of heat will be released. If the reactants in an *endothermic* reaction are doubled, then twice the amount of heat will be absorbed.

HESS'S LAW

When a reaction is composed of multiple steps, its total enthalpy change will be the sum of all the ΔH values of each of the steps. This is known as **Hess's law**. Look at the following set of reactions in the combustion of carbon to form carbon monoxide:

Step 1: $C(s) + O_2(g) \rightarrow CO_2(g)$ $\Delta H_1 = -394$ kJ

Step 2: $CO_2(g) \rightarrow CO(g) + \frac{1}{2}O_2(g)$ $\Delta H_2 = 283$ kJ

If we combine the two half reactions, we get an overall reaction of:

$$C(s) + \frac{1}{2}O_2(g) \rightarrow CO(g)$$

The total ΔH_f° for the overall reaction is:

$$\Delta H_f^\circ = -394 \text{ kJ/mol} + 283 \text{ kJ/mol} = -111 \text{ kJ/mol}$$

Remember that the total change in enthalpy is also equal to the enthalpy of the reactants subtracted from the enthalpy of the products. For example, consider the following reaction:

$$C_6H_{12}O_6 + 6O_2 \rightarrow 6CO_2 + 6H_2O$$

(-1267) $6(0)$ $6(-393.5) + 6(-258.8)$

$$\Delta H = H_p - H_r$$

$$\Delta H = -2808.8 \text{ kJ/mol}$$

GIBBS FREE ENERGY

We mentioned that the ideal state for compounds is one of low energy and high entropy. These two ideas can be combined into one equation that can be used to determine whether or not a reaction is spontaneous. This is known as the **Gibbs free energy equation**, and is shown below.

$$\Delta G = \Delta H - T\Delta S$$

T = temperature in Kelvin

If ΔG is negative, then the reaction is spontaneous.

If ΔG is positive, then the reaction is not spontaneous.

If ΔG is equal to zero, then the reaction is at equilibrium.

The free energy (ΔG) for a reaction can be calculated by subtracting the free energies of formation (ΔG_f) of the reactants from the standard free energies of formation of the products.

$$\Delta G_f = \Delta G_{fp} - \Delta G_{fr}$$

Generally speaking, systems tend to move toward lower enthalpy and higher entropy. Since these are contradictory states, the spontaneous process will result in decreasing enthalpy or increasing entropy.

We notice in the Gibbs free energy formula that ΔG also depends on temperature.

$$\Delta G = \Delta H - T\Delta S$$

When the temperature is very low, then $T\Delta S$ is insignificant and the change in enthalpy basically determines the spontaneity of the reaction. If the temperature is high, then $T\Delta S$ is quite significant and the change in entropy plays a big part in determining the spontaneity of the reaction. The following chart summarizes possibilities for ΔG.

ΔH	ΔS	ΔG
−	+	− (spontaneous)
+	−	+ (not spontaneous)
+	+	− (only if temperature is high)
−	−	− (only if temperature is low)

STANDARD STATES

The values of energy, enthalpy, and entropy provided on your test will most likely represent the **standard state conditions**. These are the values listed in thermodynamic reference tables. The standard state thermodynamic conditions assume the following:

- Gases are at a pressure of 1 atm.
- The temperature is at 25 °C (298 K).
- Solutions are at a concentration of 1 *M*.

The standard state is depicted by the degree symbol (°).

ENERGY DIAGRAMS

A **reaction energy diagram** reveals how enthalpy changes in the course a reaction. There are two types of diagrams: one for exothermic reactions and one for endothermic reactions.

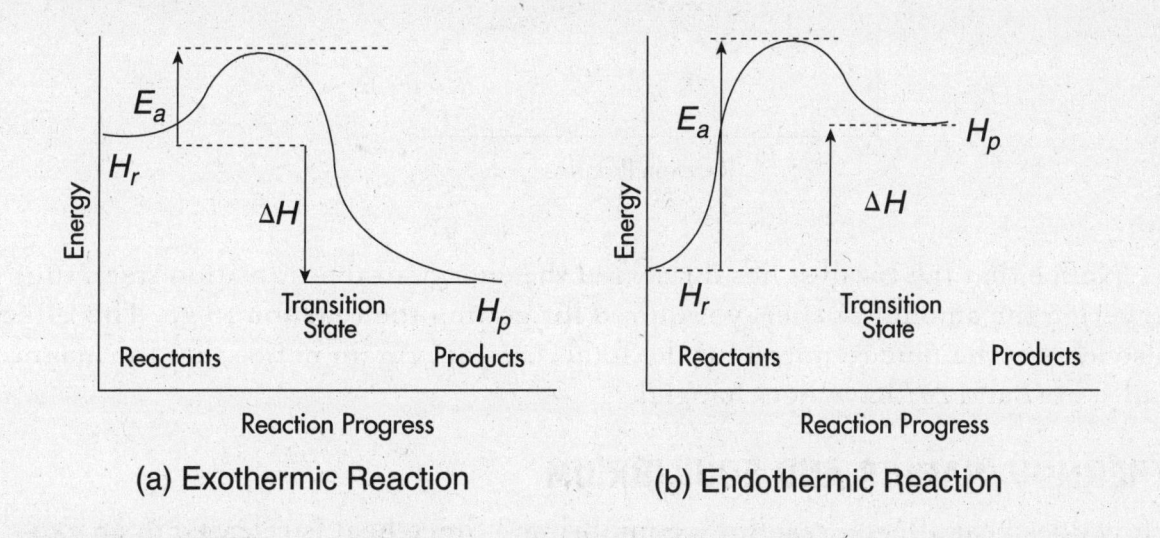

(a) Exothermic Reaction

(b) Endothermic Reaction

The diagram starts off with the enthalpy value of the reactants, and the energy changes as the reaction progresses. The peak in the graph represents the activated complex, or **transition state** of the reaction—this is the point at which the substances have acquired enough energy to make the reaction "go." The reaction is very unstable at this point because of the high energy state. The amount of energy required to bring the reaction to this state is known as the **activation energy** (E_a).

The exothermic diagram shows that the energy of the products is less than the energy of the initial products, while the endothermic reaction shows that the energy of the products is greater than the energy of the reactants. If you were to reverse the direction of the exothermic diagram, you would have the endothermic reaction. The values of H and $\Delta H°$ will be the same, except that the ΔH values will have opposite signs. The only value that will be different is the activation energy; more energy is required to drive the endothermic reaction than to drive the exothermic reaction.

CATALYSTS

How do catalysts affect an energy diagram? Earlier we talked about how catalysts are used to speed up reactions. The addition of a catalyst alters the path of the reaction but not the reactants and products themselves.

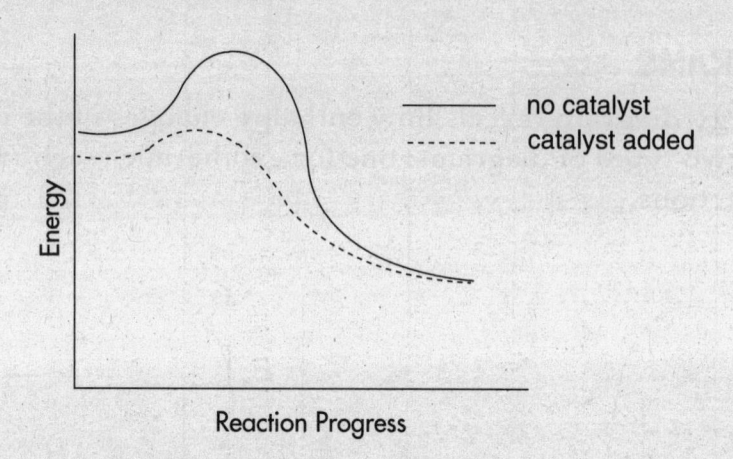

Notice that the catalyst has decreased the energy of the transition state, thus lowering the amount of energy required for getting the reaction to go. This effect also lessens the time required for the total reaction. Again notice that the magnitudes of H and ΔH have not changed.

THERMODYNAMICS AND EQUILIBRIUM

How does heat affect a reaction's equilibrium? Since heat is released in an exothermic reaction, it can be considered a product. And since heat is absorbed during an endothermic reaction, it can be considered as a reactant.

$A + B \rightarrow C + D +$ heat   Exothermic reaction $(-\Delta H)$

$A + B +$ heat $\rightarrow C + D$ Endothermic reaction $(+\Delta H)$

LeChatlier's principle tells us that exothermic reactions will shift in the reverse direction if we increase its temperature, while endothermic reactions will proceed in the forward direction.

How does energy affect equilibrium? The following equation shows the relationship of $\Delta G°$ and ΔG:

$$\Delta G = \Delta G° + -2.303\, RT \log K$$

Earlier we mentioned that at equilibrium, ΔG equals zero. Therefore,

$$\Delta G° = -2.303\, RT \log K$$

R = the gas constant (8.31 J/mol × K)

T = absolute temperature (K)

K = equilibrium constant

Under standard conditions, when $\Delta G°$ is negative, the equilibrium constant (K) is greater than one. Therefore, when a reaction is spontaneous, it moves in the forward direction and more product is formed. When $\Delta G°$ is positive, the equilibrium constant (K) is less than one. Therefore, when a reaction is not spontaneous, the reaction moves in the reverse direction and more reactants are formed.

THERMODYNAMICS

1. Adding a catalyst to a chemical reaction will result in which of the following?

   A. The enthalpy change (ΔH) will increase

   B. The enthalpy change (ΔH) will decrease

   C. The entropy change (ΔS) will decrease

   D. The activation energy will decrease

Answer:

D. Adding a catalyst to a reaction will speed up the reaction by lowering the activation energy having no effect on the change in entropy or enthalpy.

2. If an exothermic reaction is taking place and temperature is increased, which of the following will occur?

   A. The rate of the forward reaction will increase and the rate of the reverse reaction will increase.

   B. The rate of the forward reaction will decrease and the rate of the reverse reaction will increase.

   C. The rate of the forward reaction will increase and the rate of the reverse reaction will decrease.

   D. The rate of the forward reaction will decrease and the rate of the reverse reaction will decrease.

Answer:

B. According to LeChatelier's principle a reaction will shift in the direction that will relieve the stress that has been placed on the system. With an exothermic reaction heat is released by the reaction, thus acting as a product of the reaction. Therefore, if temperature is increased the reverse reaction would increase (and the forward reaction would decrease) in order to consume the excess temperature that has been introduced to the reaction.

3.

Bond	Average Bond Dissociation Energy (kJ/mol)
C—H	415
O=O	495
C=O	799
O—H	463

Using the bond dissociation energies listed above, what is the ΔH for the reaction of methane with oxygen?

$$CH_4(g) + 2O_2(g) \rightarrow CO_2(g) + 2H_2O(l)$$

A. −1726 kJ

B. −800 kJ

C. 800 kJ

D. 1726 kJ

Answer:

ΔH = sum of energies of bonds broken − sum of energies of bonds formed

$$\Delta H = [(4)(415) + (2)(495)] - [(2)(799) + (4)(463)]$$

$$\Delta H = -800 \text{ kJ/mol}$$

4. If a reaction is spontaneous which of the following must be correct?

A. The reaction is exothermic

B. The reaction is endothermic

C. ΔS is negative

D. ΔG is negative

Answer:

D. ΔG is defined as the change in free energy.

$$\Delta G = \Delta H - T\Delta S$$

All reactions that experience a decrease in free energy are spontaneous ($\Delta G < 0$) Endothermic reactions have a positive ΔH meaning that energy is absorbed. An endothermic reaction is generally non spontaneous. Although energy is released with an exothermic reaction, if the ΔS value is large enough an exothermic reaction CAN be nonspontaneous. Remember the question asks for which statement MUST be true. A negative or positive ΔS can be spontaneous or nonspontaneous depending on the ΔH value.

5.

Potential Energy

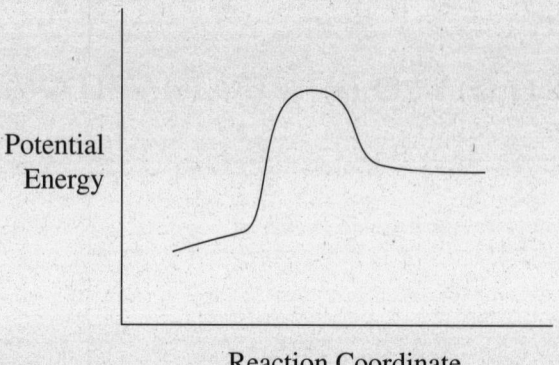

Reaction Coordinate

Which of the following statements is correct regarding the reaction depicted above?

A. The reaction is exothermic since the reactants are at a higher energy level than the products.

B. The reaction is exothermic since the reactants are at a lower energy level than the products.

C. The reaction is endothermic since the reactants are at a lower energy state than the products.

D. The reaction is endothermic since the reactants are at a higher energy state than the products.

Answer:

C. In the diagram the products are at a higher energy state than the reactants. In order to increase the energy state in the reaction, energy must have been absorbed. When energy is absorbed by the reaction, the reaction is endothermic.

Chapter 12

ELECTROCHEMISTRY

In an earlier chapter, we talked about oxidation-reduction (redox) reactions. If we wanted to find out if a redox reaction was spontaneous, we would need to know each half reaction's potential, or **voltage** ($E°$), measured in volts (V). Don't worry, you do not have to memorize any of the voltages; if you need them, they will be provided to you on the GSE.

When the reduction potential is negative, the reaction is not spontaneous. For example, let's look at the reduction of nickel:

$$Ni^{2+} + 2e^- \rightarrow Ni(s) \quad E° = -0.25 \text{ V}$$

A negative reduction potential tells us that nickel will not be spontaneously reduced.

The opposite reaction represents the oxidation of nickel:

$$Ni(s) \rightarrow Ni^{2+} + 2e^- \quad E° = +0.25 \text{ V}$$

When the potential is positive, a reaction is spontaneous. Thus the positive voltage listed above tells us that nickel is spontaneously oxidized.

Notice that the voltage for the oxidation of nickel has the same magnitude as the reduction potential, but the opposite sign. Charts will list the half reactions as reduction potentials. If you need the oxidation potential, just flip the reaction *but don't forget to change the sign of the reduction potential!*

In order to determine the reaction potential for an entire redox reaction, add the potentials for both the reduction reaction and the oxidation reaction. For example, look at the following equation:

$$Zn^{2+}(aq) + 2Fe^{2+}(aq) \rightarrow Zn(s) + 2Fe^{3+}(aq)$$

First we need to split this redox reaction into two half reactions.

$$Zn^{2+}(aq) + 2e^- \rightarrow Zn(s)$$

$$2Fe^{2+}(aq) \rightarrow 2Fe^{3+}(aq) + 2e^-$$

If we look at a voltage reference table, we can find the following information:

$$Zn^{2+}(aq) + 2e^- \rightarrow Zn(s) \quad E° = -0.76 \text{ V}$$

$$Fe^{3+}(aq) + e^- \rightarrow Fe^{2+}(aq) \quad E° = 0.77 \text{ V}$$

Since iron is oxidized in this redox reaction, we need to change the sign of the reduction potential. Therefore, the total potential of our redox reaction is:

$$E° = E°_{red} + E°_{ox}$$

$$= -0.76 \text{ V} + -0.77 \text{ V} = -1.53 \text{ V}$$

Since the total potential is negative, this original redox reaction is not spontaneous.

Notice that we did not multiply the iron's potential by any coefficient. Even if a half reaction is multiplied by a coefficient in order to obtain a balanced equation, never multiply the potential by a coefficient.

According to thermodynamics, the reverse of a nonspontaneous reaction is spontaneous, and vice versa. This means that the potential of the reverse reaction of a redox equation would make it spontaneous.

$$Zn(s) + 2Fe^{3+}(aq) \rightarrow Zn^{2+}(aq) + 2Fe^{2+}(aq) \quad E° = +1.53 \text{ V}$$

GALVANIC CELLS

A **galvanic cell** (or voltaic cell) is an electrochemical cell in which the transfer of electrons during a spontaneous redox reaction produces an electric current. The following illustration represents a galvanic cell:

Galvanic Cell

e⁻ flow →

Anode (−)
- e⁻ rich electrode
- Oxidation reaction supplies electrons
- Electrolyte A

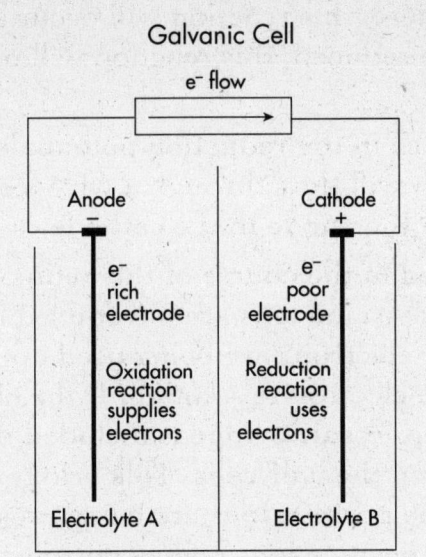

Cathode (+)
- e⁻ poor electrode
- Reduction reaction uses electrons
- Electrolyte B

spontaneous oxidation-reduction reaction—source of energy

In the galvanic cell, the half reactions occur in two different chambers known as half-cells. The electrons are transferred from the oxidation half-cell (via an external wire) to the reduction half-cell.

It is important to understand how the galvanic cell works. The reaction in each half-cell takes place at an **electrode**. The electrode at which the oxidation reaction takes place is known as the **anode** and the electrode at which the reduction reaction takes place is the **cathode**. The following mnemonic device is used to remember the reaction process:

AN OX (anode – oxidation)

RED CAT (cathode – reduction)

The solution in each half-cell is comprised of dissolved ions and is capable of conducting electricity. These solutions are **electrolytes**. Each cell contains a different electrolyte.

Look at the following redox reaction:

$$Zn(s) + Cu^{2+}(aq) \rightarrow Zn^{2+}(aq) + Cu(s) \quad E° = 1.10 \text{ V}$$

Half reactions:

$$Zn(s) \rightarrow Zn^{2+}(aq) + 2e^- \quad E° = 0.76 \text{ V}$$

$$Cu^{2+}(aq) + 2e^- \rightarrow Cu(s) \quad E° = 0.34 \text{ V}$$

Since zinc is oxidized here, this reaction will occur at the negatively charged anode, and since copper is reduced, this reaction will occur at the positively charged cathode.

Because of the difference in the reduction potentials, the electrons that are produced during oxidation will flow through a wire from one half-cell into another. Electrons flow from the anode to the cathode.

Many ions are produced in the course of this redox reaction. As electrons flow from the anode, positive zinc ions are introduced into the electrolyte solution. In the other half-cell, electrons are being used up to remove the positively charged ion Cu^{2+} from the electrolyte solution. If the charges continued to build up, the reaction would stop. A **salt bridge** is included in the setup to maintain electrical neutrality between the half-cells. This bridge will contribute negative ions to the anode where the positive ions are being created, and positive ions to the cathode to replace the positive ions being consumed. Therefore, from the salt bridge: Anions are contributed to the anode, and cations are contributed to the cathode.

The voltage of the cell under standard conditions is equal to the voltage (or reduction potential) of the redox reaction. However, the following **Nernst equation** calculates the total voltage of a cell under nonstandard conditions:

$$E_{cell} = E_{cell}° - \left(\frac{RT}{nF}\right)\ln Q$$

E_{cell} = cell voltage (nonstandard conditions)

$E°_{cell}$ = cell voltage (standard conditions)

R = the gas constant, $8.31 \left(\frac{V \cdot coul}{mol \cdot K}\right)$

T = absolute temperature (K)

n = number of moles of electrons transferred in the reaction

F = Faraday's constant (96,500 coulombs/mol)

Q = Reaction quotient (an expression that is the same as the equilibrium expression, except that it considers the initial concentrations instead of the equilibrium concentrations)

The Nernst equation can be written in the following way at a standard temperature of 25 °C:

$$E_{cell} = E_{cell}° - \frac{0.0592}{n}\log Q$$

This equation reveals that voltage decreases as the concentrations of reactants decrease. After all the reactants are gone, the reaction will stop, since the reverse direction of a spontaneous reaction is nonspontaneous.

ELECTROLYTIC CELLS

There is another type of cell that you need to be familiar with for the Chemistry GSE. It is called an **electrolytic cell**. We reviewed how a galvanic cell uses the energy produced during a redox reaction to generate a current. An electrolytic cell takes an electric current from an outside source, such as a battery, to drive a nonspontaneous chemical reaction.

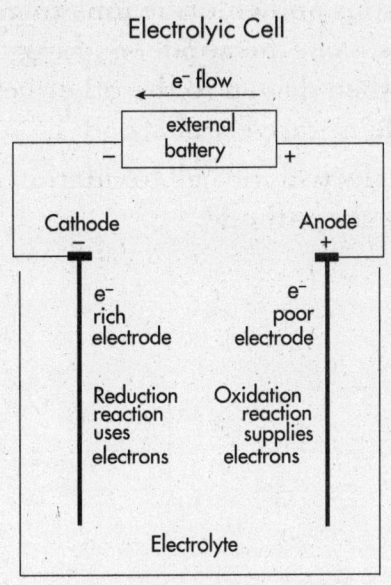

Electrolyic Cell

Driven oxidation-reduction reaction—a user of energy

Many properties of an electrolytic cell make it different from a galvanic cell. The following chart outlines these differences.

GALVANIC CELL	ELECTROLYTIC CELL
spontaneous reaction	nonspontaneous reaction
produces a current	consumes a current
cathode— positive electrode	anode— positive electrode
anode— negative electrode	cathode— negative electrode
electrons flow from anode to cathode	electrons flow from cathode to anode

ELECTROCHEMISTRY ◆ 123

How does the electrolytic cell work? Take a look at the electrolysis of molten NaCl, represented by the following reaction:

$$2\ Na^+(molten) + 2\ Cl^-(molten) \rightarrow 2Na(l) + Cl_2(g) \quad E° = -4.07\ V$$

Half reactions:

$$2Cl^-\ (molten) \rightarrow Cl_2(g) + 2e^- \quad E° = -1.36\ V$$

$$2Na^+\ (molten) + 2e^- \rightarrow 2Na(l) \quad E° = -2.71\ V$$

Notice that the overall potential of the reaction is negative ($E° = -4.07\ V$), which means that the reaction is not spontaneous. The external battery will force sodium ions to accept electrons and chloride ions to give up electrons. The battery will pull these electrons from the anode, making the anode positively charged. The electrons are then driven to the other cell and are forced onto the cathode, making the cathode negatively charged.

Keep in mind that in an electrolytic cell, oxidation still occurs at the anode and reduction still occurs at the cathode.

ELECTROCHEMISTRY

1. If the oxidation state of an atom changes from 0 to –2 during an oxidation-reduction reaction, which of the following is correct?

   A. The atom is oxidized and serves as a reducing agent

   B. The atom is reduced and serves as a reducing agent

   C. The atom is oxidized and serves as an oxidizing agent

   D. The atom is reduced and serves as an oxidizing agent

Answer:

D. Since the atom's oxidation state changed from 0 to –2, the atom has taken on two electrons. When an atom gains electrons it is being reduced and it is taking electrons away from another atom, which is being oxidized. An atom being reduced will cause another atom to be oxidized, thus serving as an oxidizing agent.

2. $3\ MnO_2 + 4\ Al \rightarrow 2Al_2O_3 + 3Mn$

   The reduction half reaction for the above oxidation-reduction reaction is illustrated by which of the following?

   A. $Mn^{4+} + 4\ e^- \rightarrow Mn^0$

   B. $Mn^0 \rightarrow Mn^{2+} + 2e^-$

   C. $Al^0 \rightarrow Al^{3+} + 3e^-$

   D. $Al^{3+} + 3e^- \rightarrow Al^0$

Answer:

A. During reduction an atom gains electrons. The oxidation state of Mn in MnO_2 is equal to +4, while product Mn has an oxidation state of 0. Mn therefore, accepted four electrons in order to be reduced from +4 to 0.

3. Which of the following statements reveals how electrons will flow during a reaction in a galvanic cell?

   A. From cathode to the anode

   B. From the anode to the cathode

   C. From the anode to the cathode then back to the anode

   D. There is no flow of electrons during a reaction in a galvanic cell

Answer:

B. In a galvanic cell electrons flow from the cell where oxidation occurs freeing up electrons to the cell where the reduction reaction takes place using up the supply of electrons. Oxidation will occur at the anode and reduction occurs at the cathode, therefore, electrons will flow from the anode to the cathode.

4. In an electrolytic cell, which of the following statements are true?

   I. The reaction is nonspontaneous

   II. The site of oxidation is the anode

   III. The positive electrode is the cathode

   A. I only
   B. I and II only
   C. II and III only
   D. I, II and III

Answer:

B. An electrolytic cell uses an outside source of energy to produce a chemical reaction that would not happen spontaneously. Oxidation will still occur at the anode, however the positive electrode will be the anode.

5. $Fe^{2+} + Cu \rightarrow Fe + Cu^{2+}$

   Use the following reduction potentials to determine the reaction potential for the reaction of iron with copper as depicted above.

   $$Cu^{2+} + 2e^- \rightarrow Cu \quad E = +0.3V$$

   $$Fe^{2+} + 2e^- \rightarrow Fe \quad E = -0.4V$$

   A. −0.7V
   B. −0.1V
   C. +0.1V
   D. +0.7V

Answer:

A. In the reaction above Fe^{2+} is being reduced and Cu is being oxidized. Therefore, we add the reduction potential for Fe^{2+} (−0.4V) to the oxidation potential for Cu (−0.3V) [reverse of the reduction potential listed] to obtain −0.7V.

Chapter 13

NUCLEAR DECAY

The nucleus of an atom is held together by forces between the protons and neutrons known as **nuclear force**. Some nuclei are very unstable and try to become more stable by altering their content of protons and neutrons in a process called **radioactive** or **nuclear decay**.

TYPES OF NUCLEAR DECAY

Alpha Decay
Alpha decay occurs when an unstable nucleus emits a particle that has the same number of protons and neutrons as a helium atom. This **alpha particle** is composed of two protons and two neutrons and has a mass number of four.

$$_{2}^{4}\alpha = {}_{2}^{4}\text{He}$$

When a nucleus emits an alpha particle, the following occurs:
1. The atomic number decreases by two.
2. The mass number decreases by four.

For example:

$$^{238}_{92}U = ^{4}_{2}\alpha + ^{234}_{90}Th$$

Beta Decay

Beta decay occurs when an unstable nucleus converts a neutron into a proton and a beta particle. The beta particle is identical to an electron and is emitted in the process.

$$^{1}_{0}n \rightarrow ^{0}_{-1}\beta + ^{1}_{1}p$$

When a nucleus undergoes beta decay, the following occurs:
1. The atomic number increases by one.
2. The mass number stays the same.

For example:

$$^{14}_{6}C \rightarrow ^{0}_{-1}\beta + ^{14}_{7}N$$

Positron Emission

Positron emission occurs when an unstable nucleus converts a proton into a neutron and a positron. The positron particle has the same mass as an electron but a positive charge; it is emitted during this process.

$$^{1}_{1}p \rightarrow ^{0}_{1}\beta + ^{1}_{0}n$$

When a nucleus undergoes positron emission, the following occurs:
1. The atomic number decreases by one.
2. The mass number stays the same.

For example:

$$^{8}_{5}B \rightarrow ^{0}_{1}\beta + ^{8}_{4}Be$$

Electron Capture

Some unstable nuclei convert a proton into a neutron by capturing an electron from the first energy shell.

$$^{0}_{-1}e + ^{1}_{1}p \rightarrow ^{1}_{0}n$$

During electron capture, the following occurs:
1. The atomic number decreases by one.
2. The mass number stays the same.

Gamma Rays

Gamma rays are a form of electromagnetic radiation that are often emitted together with alpha particles, beta particles, and positrons. Gamma rays have no mass or protons.

$^0_0\gamma$ Gamma rays contain the highest amount of energy.

BINDING ENERGY AND MASS DEFECT

The total mass of a nucleus is actually less than the sum of the masses of its protons and neutrons. When neutrons and protons converge to form the nucleus, a portion of the total mass is lost in the form of energy. This difference in mass is known as the **mass defect**.

When the process is reversed, the same amount of energy is required to break the nucleus into its individual protons and neutrons. This is known as **binding energy**. The following equation from Albert Einstein depicts the relationship between mass and energy:

$E = mc^2$

E = energy (Joules)

m = mass (kg)

c = speed of light (3×10^8 m/sec)

Since c^2 is a very large number, a very small change in mass will result in a very large change in energy.

NUCLEAR STABILITY

Nuclei of unstable isotopes will undergo decay in order to become stable. You can use the following guidelines in order to predict the type of decay that they will undergo:

- If the mass number is larger than twice the atomic number, the nucleus will attempt to acquire protons and lose neutrons through beta decay.

- If the mass number is less than twice the atomic weight, the nucleus will attempt to lose protons and gain neutrons through positron emission or electron capture.

- Alpha decay occurs most often in unstable nuclei that have atomic numbers greater than sixty.

HALF-LIFE

The time that it takes for half of a radioactive substance to decay is known as its **half-life**. For example, what if we started with 320 g of a radioactive substance that has a half-life of 5 years, how many years will it take for only 10 g of that substance to remain?

SAMPLE (g)	TIME (years)
160	5
80	10
40	15
20	20
10	25

After 25 years, our 320 g sample will be reduced to 10 g, if the half-life is 5 years.

NUCLEAR DECAY

1. If a 340 g sample of Substance Z has a half life of 7 years, how many years will be needed to reduce the sample size to 21.25g?

   A. 7 years

   B. 14 years

   C. 21 years

   D. 28 years

   Answer:

   D. Draw a chart to help you out.

Number of years	Sample Size (g)
0	340
7	170
14	85
21	42.5
28	21.25

2. Which of the following nuclides is a result of carbon-14 decaying by emitting a beta particle?

   A. boron-13

   B. carbon-15

   C. nitrogen-14

   D. nitrogen-15

   Answer:

   C. When a nucleus undergoes beta emission, it converts a neutron into a proton and an electron. The atomic number will increase by one and the mass number stays the same. Increasing carbon by one atomic number results in nitrogen with the same mass number of 14.

NUCLEAR DECAY ◆ 131

3. Which of the following decay process will result in the number of protons in a nuclide decreasing while the mass number stays the same?

   I. beta decay

   II. positron emission

   III. electron capture

A. I only

B. II only

C. II and III only

D. I, II and III

Answer:

C. I. During beta decay a nucleus converts a neutron into a proton and an electron. This results in the atomic number increasing by one. Since I is incorrect cross off answer choices A and D. (Notice that II now appears in both of the remaining answer choices which means II has to be correct). III. During electron capture a nucleus converts a proton to a neutron by capturing an electron from its electron shell. This again results in the atomic number to decrease by one while mass number remains constant.

4. If $^{12}_{5}B$ nuclide decays to $^{8}_{3}Li$ which of the following process must have occurred?

A. Alpha decay

B. Beta decay

C. Electron capture

D. Positron Emission

Answer:

A. In alpha decay, a nuclide emits an alpha particle which consists of 2 protons and 2 neutrons. This will result in the atomic number decreasing by 2 and the mass number decreasing by 4.

$$^{12}_{5}B \rightarrow\ ^{8}_{3}Li +\ ^{4}_{2}\alpha$$

$$^{12}_{5}B \rightarrow\ ^{8}_{3}Li +\ ^{4}_{2}\alpha$$

5. A radioactive substance had decayed to 12.5% of its original quantity after 60 minutes. What is the half life of this substance?

A. 15 minutes

B. 30 minutes

C. 60 minutes

D. 75 minutes

Answer:

D. Draw a chart to help you out.

Half lives	Time	Sample %
0	0	10%
1	20	50%
2	40	25%
3	60	12.5%

It takes three half lives to reduce the sample quantity to 12.5 % of the original amount. Therefore, each half live must be equal to 20 minutes.

Chapter 14

ORGANIC CHEMISTRY AND BIOCHEMISTRY

The study of carbon compounds is referred to as **organic chemistry**. Carbon can form up to four bonds that are usually covalent with minimal polarity.

THE HYDROCARBONS

Hydrocarbons are the simplest organic compounds—they contain only carbon and hydrogen. **Alkanes** contain only single bonds. They are also known as *saturated hydrocarbons*, since each carbon atom bonds to as many other atoms as possible.

Alkane (C_nH_{2n+2}...)	Formula
Methane	CH_4
Ethane	C_2H_6
Propane	C_3H_8
Butane	C_4H_{10}
Pentane	C_5H_{12}

These compounds are also known as hydrocarbon chains.

Methane   Ethane   Propane

$$H-\underset{\underset{H}{|}}{\overset{\overset{H}{|}}{C}}-H \qquad H-\underset{\underset{H}{|}}{\overset{\overset{H}{|}}{C}}-\underset{\underset{H}{|}}{\overset{\overset{H}{|}}{C}}-H \qquad H-\underset{\underset{H}{|}}{\overset{\overset{H}{|}}{C}}-\underset{\underset{H}{|}}{\overset{\overset{H}{|}}{C}}-\underset{\underset{H}{|}}{\overset{\overset{H}{|}}{C}}-H$$

Alkenes contain double bonds and are known as unsaturated hydrocarbons.

Alkene (C_nH_{2n}...)	Formula
Ethylene	C_2H_4
Propene	C_3H_6
Butene	C_4H_8
Pentene	C_5H_{10}

Ethylene   Propene

$$\underset{H}{\overset{H}{>}}C=C\underset{H}{\overset{H}{<}} \qquad H-\underset{\underset{H}{|}}{\overset{\overset{H}{|}}{C}}-\underset{\underset{H}{|}}{C}=C\underset{H}{\overset{H}{<}}$$

Alkynes contain triple bonds (they are also known as *unsaturated hydrocarbons*).

Alkyne (C_nH_{2n-2}...)	Formula
Ethyne	C_2H_2
Propyne	C_3H_4
Butyne	C_4H_6
Pentyne	C_5H_8

Ethyene   Propyne

$$H-C\equiv C-H \qquad H-\underset{\underset{H}{|}}{\overset{\overset{H}{|}}{C}}-C\equiv C-H$$

Many hydrocarbons form hydrocarbon rings instead of chains. The **aromatic hydrocarbon** is one of the most important classes of these compounds. The simplest form of an aromatic hydrocarbon is benzene (C_6H_6).

Benzene

The benzene ring is sometimes represented by the following structure:

Abbreviated Benzene

FUNCTIONAL GROUPS

The following **functional groups** give certain compounds specific chemical properties when present:

Alcohols are organic compounds that contain a hydroxyl group (–OH) in the place of a hydrogen atom. Alcohols are generally polar and more soluble in water than compounds that do not contain a hydroxyl group.

Alcohol	Formula
Methanol	CH_3OH
Ethanol	C_2H_5OH
Propanol	C_3H_7OH
Butanol	C_4H_9OH
Pentanol	$C_5H_{11}OH$

Methanol

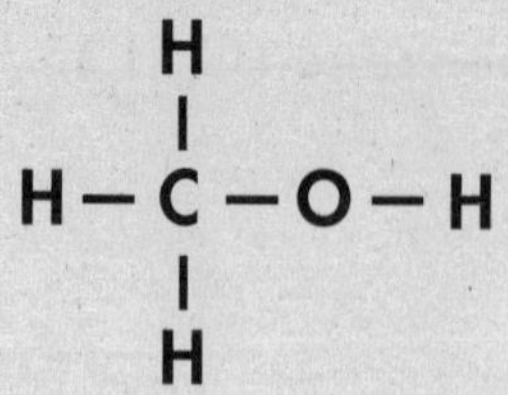

```
      H
      |
  H - C - O - H
      |
      H
```

Ethanol

```
      H   H
      |   |
  H - C - C - O - H
      |   |
      H   H
```

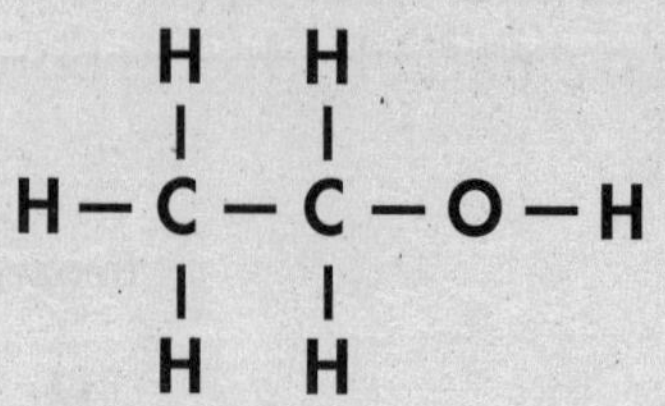

Organic acids are compounds that contain a carboxyl group (–COOH) in the place of a hydrogen atom.

Organic Acid	Formula
Formic acid	HCOOH
Acetic acid	CH$_3$COOH
Butyric acid	C$_3$H$_7$COOH

Formic acid

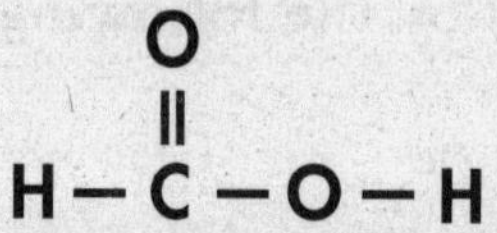

```
      O
      ||
  H - C - O - H
```

Acetic acid

```
      H   O
      |   ||
  H - C - C - O - H
      |
      H
```

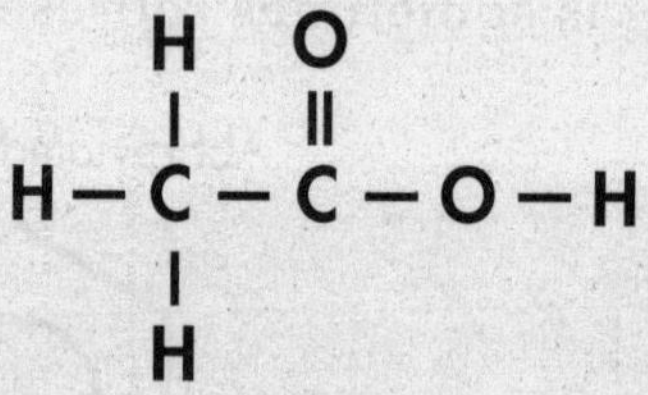

Halides are organic compounds that contain a halide (F, Cl, Br, I) in the place of one of more hydrogen atoms.

Halide	Formula
Chloromethane	CH$_3$Cl
Chloroethane	C$_2$H$_5$Cl
Chloropropane	C$_3$H$_7$Cl

Chloromethane

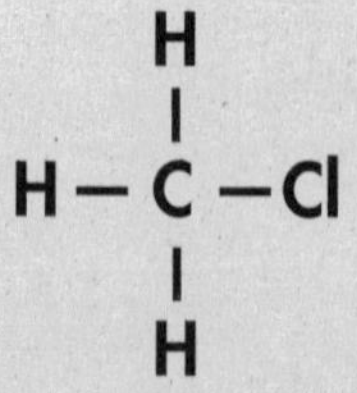

```
      H
      |
  H - C - Cl
      |
      H
```

Chloroethane

```
      H   H
      |   |
  H - C - C - Cl
      |   |
      H   H
```

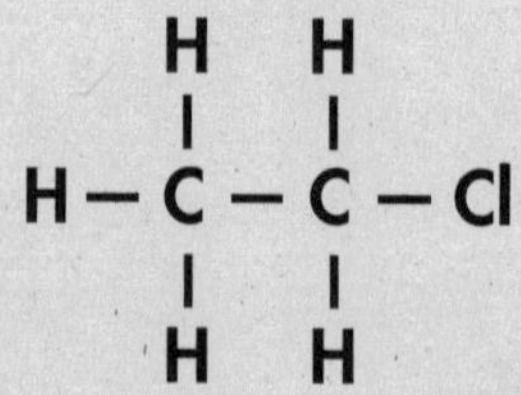

Amines are organic compounds that contain an amino group (NH$_2$) in place of a hydrogen atom.

138 ◆ CRACKING THE GOLDEN STATE EXAMINATION: CHEMISTRY

Amine	Formula
Methylamine	CH_3NH_2
Ethylamine	$C_2H_5NH_2$

Methylamine

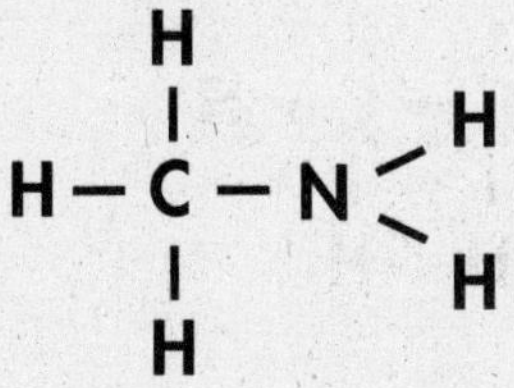

```
      H
      |      H
  H — C — N ⟨
      |      H
      H
```

Ethylamine

```
      H   H
      |   |      H
  H — C — C — N ⟨
      |   |      H
      H   H
```

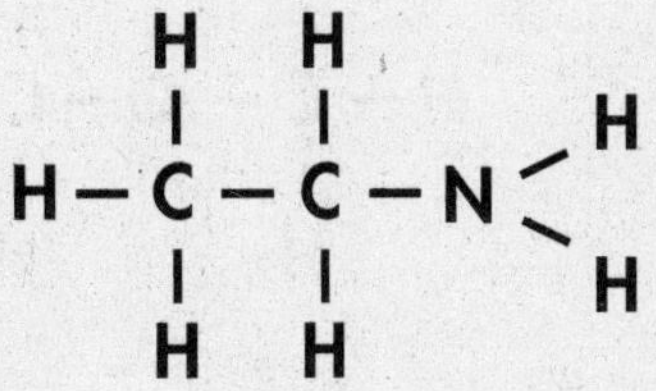

Aldehydes are organic compounds that contain a carbonyl group (C=O) connected to at least one hydrogen atom.

Aldehyde	Formula
Methanal (formaldehyde)	CH_2O
Ethanal (acetaldehyde)	CH_3CHO

Methanal

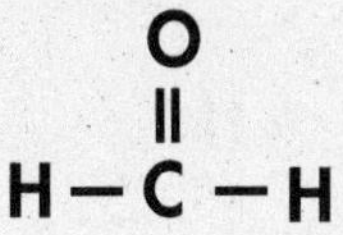

```
      O
      ‖
  H — C — H
```

Ethanal

```
      H   O
      |   ‖
  H — C — C — H
      |
      H
```

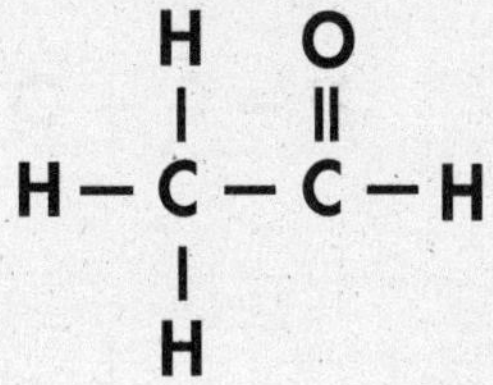

Ketones are similar to aldehydes. They are organic compounds that contain a carbonyl group (C=O), but the carbonyl group is not connected to any hydrogen atoms. Rather, it is connect to other groups, such as –CH_3 in acetone.

Ketone	Formula
Acetone	CH_3COCH_3

Acetone

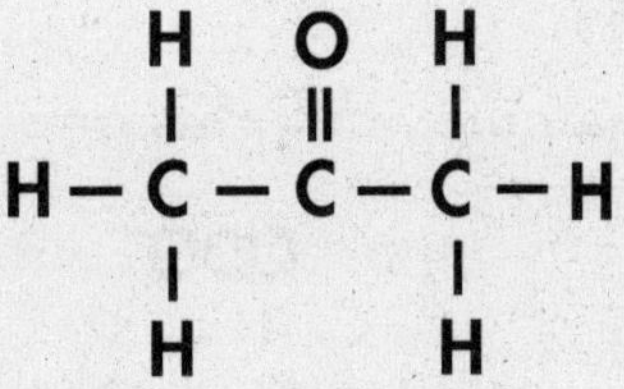

```
      H   O   H
      |   ‖   |
  H — C — C — C — H
      |       |
      H       H
```

Ethers are compounds in which an oxygen atom links two other groups in a hydrocarbon chain.

Ether	Formula
Methylpropylether	CH$_3$OCH$_2$CH$_2$CH$_3$

Methylpropylether

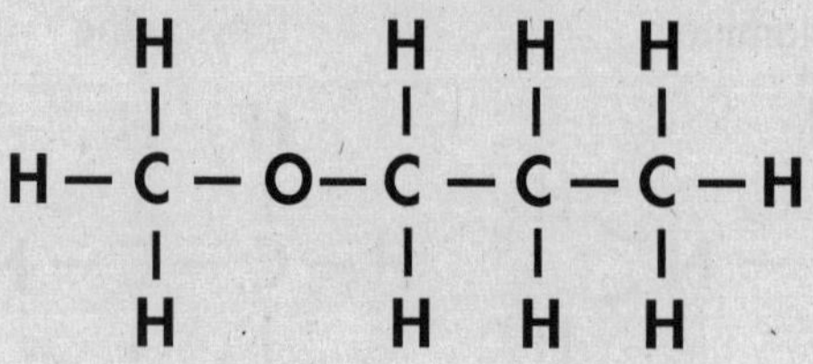

```
    H       H   H   H
    |       |   |   |
H — C — O — C — C — C — H
    |       |   |   |
    H       H   H   H
```

Esters are compounds where an ester group (–COO–) links two other groups in a hydrocarbon chain.

Ester	Formula
Ethyl formate	HCOOCH$_2$CH$_3$
Methyl butyrate	CH$_3$CH$_2$CH$_2$COOCH$_3$

Ethyl formate

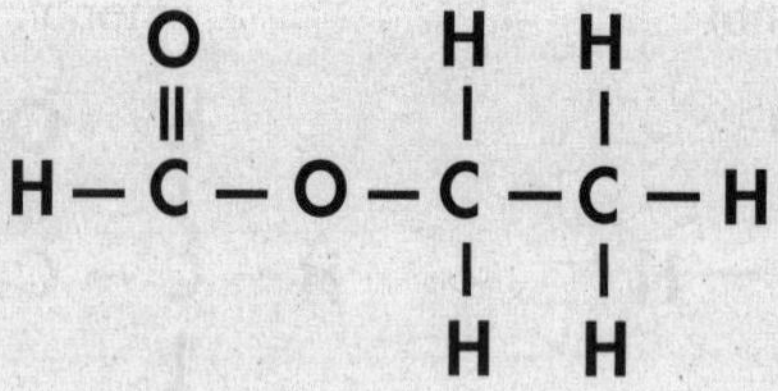

```
    O       H   H
    ||      |   |
H — C — O — C — C — H
            |   |
            H   H
```

BIOLOGICAL MOLECULES

There are four major classifications of biological molecules you should be familiar with for the chemistry GSE: proteins, carbohydrates, nucleic acids, and lipids.

Proteins

Proteins are biological polymers comprised of amino acids. An **amino acid** contains an amino group with an additional proton (NH^{3+}) and a carboxyl group that has been ionized (COO–).

Here is an example of an amino acid, this one is called glycine—it's the simplest amino acid.

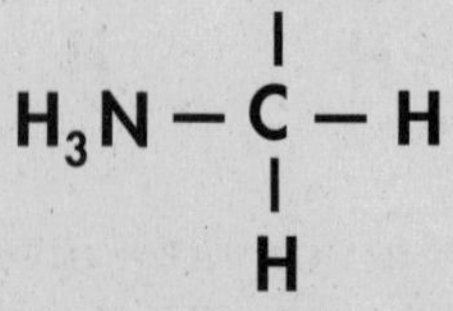

```
            COO–
            |
    H$_3$N — C — H
            |
            H
```

Amino acids are the *building blocks of proteins*. There are 20 amino acids that are collectively known as the common amino acids, but you certainly won't need to memorize what they are, or their structures, for the exam!

These amino acids differ by their side chains (their **R groups**). The side chains determine the protein's properties. Nine of these amino acids are nonpolar (because their R groups are nonpolar). The other eleven are polar amino acids. Their R groups allow them to form hydrogen bonds with other amino acids. Amino acids are linked together by **peptide bonds** to form proteins.

Carbohydrates

Carbohydrates are polyhydroxyl aldehydes or ketones. They can also be substances that yield polyhydroxyl aldehydes or ketones when they react with water.

Carbohydrates can be broken into three main categories: monosaccharides, oligosaccharides and polysaccharides. **Monosaccharides** are the simplest sugars; they contain three to nine carbon atoms. Most of these carbon atoms contain a hydroxyl group, while one is part of the carbonyl group. Since one carbon bonds to four different groups, there are two isomers for each monosaccharide, but we just show one isomer of each structure in the diagrams below. **Isomers** are molecules with the same molecular formula, but different arrangements of atoms and different chemical properties. Be familiar with the following monosaccharides:

glucose

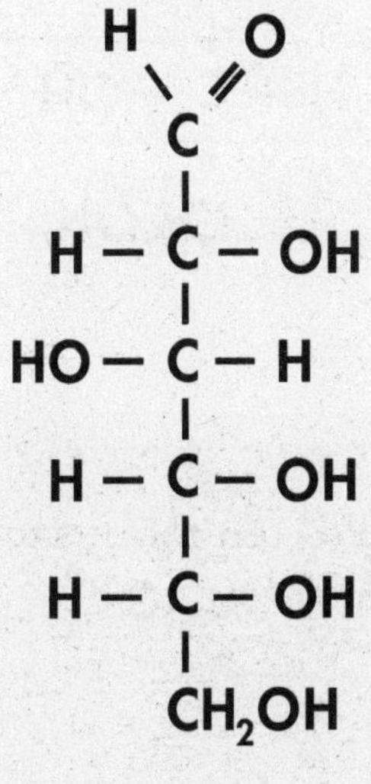

H   O
 \\ //
  C
  |
H—C—OH
  |
HO—C—H
  |
H—C—OH
  |
H—C—OH
  |
 CH$_2$OH

Fructose

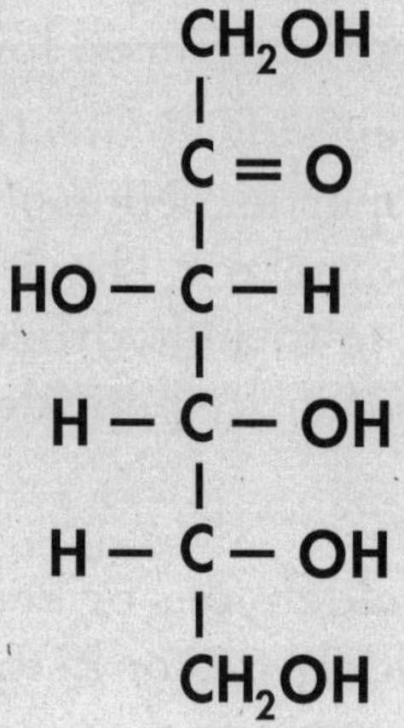

$$\begin{array}{c} CH_2OH \\ | \\ C=O \\ | \\ HO-C-H \\ | \\ H-C-OH \\ | \\ H-C-OH \\ | \\ CH_2OH \end{array}$$

Ribose

$$\begin{array}{c} HO \\ \diagdown\diagup \\ C \\ | \\ H-C-OH \\ | \\ H-C-OH \\ | \\ H-C-OH \\ | \\ CH_2OH \end{array}$$

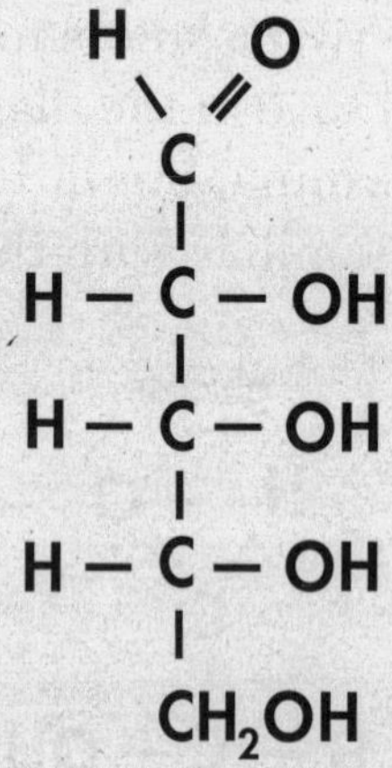

Oligosaccharides contain two to ten monosaccharides, and **polysaccharides** contain more than ten monosaccharides.

Nucleic Acids

Nucleic acids transport genetic information in the form of DNA (deoxyribonucleic acid) or RNA (ribonucleic acid). Nucleic acids are composed of two types of **nucleotides**: deoxyribonucleotide and ribonucleotide.

The following nucleotide structures contain an organic base, a sugar, and a phosphate group.

Nucleotides that link in a linear fashion form polynucleotides. This occurs between the hydroxyl group on the sugar of one nucleotide, and the phosphate group of the other nucleotide. Nucleic acids are formed when polynucleotides fold into specific three-dimensional shapes. In this process, the **complementary base groups** from different nucleotides form strong hydrogen bonds. The base **adenine** is complementary to **thymine**, while **cytosine** complements **guanine**.

DNA is comprised of two strands of deoxyribonucleotides that are bonded by hydrogen bonds between complementary bases. The strands are coiled together to form a **double helix**. DNA is the main constituent of chromosomes and transmits genetic information from parents to offspring.

RNA is comprised of a single-stranded polymer of ribonucleotide units and contains the bases **adenine, uracil (not thymine), cytosine**, and **guanine**. RNA plays an important role in the manufacture of proteins.

Lipids

Lipids are substances that are not soluble in water but are soluble in nonpolar organic solvents. Lipids include fats and oils, also known as **triacylglycerols (or triglycerides)**. Fats and oils have similar structures.

$$\begin{array}{c} \text{CH}_2-\text{O}-\overset{\overset{\displaystyle O}{\|}}{\text{C}}-\text{R}_1 \\ | \\ \text{CH}-\text{O}-\overset{\overset{\displaystyle O}{\|}}{\text{C}}-\text{R}_2 \\ | \\ \text{CH}_2-\text{O}-\overset{\overset{\displaystyle O}{\|}}{\text{C}}-\text{R}_3 \end{array}$$

This marks the end of the subject review for the Chemistry GSE. If you read carefully through this material, you should be ready to move on and take the four practice exams that follow. As you're taking the first couple of exams, feel free to flip back through the pages of the review to look up any information you'll need, but as you complete the last two exams, try to refrain from doing this. It will be helpful for you to practice taking exams without any notes to consult. Good luck!

Chapter 15

LABORATORY

Part II of the Chemistry GSE consists of one laboratory task that lasts 45 minutes. In this part of the examination you will independently perform a task using laboratory equipment and chemicals under safe conditions. You will be required to show an understanding of appropriate laboratory procedures; document observations and data in an accurate, detailed manner; support all analyses, calculations, and conclusions; and use scientific arguments to demonstrate knowledge of scientific methods.

SAFETY FIRST...

Safety is vital in a laboratory environment, and your laboratory grade will also depend in part on your ability to follow laboratory safety rules.

Always adhere to the following safety guidelines when working in a lab:

- Wear safety goggles at all times.

- Do not use your mouth to pipette chemicals.

- Make sure there is proper ventilation in the laboratory and always work with hazardous chemicals under a laboratory ventilation hood.

- Always add acid to water when diluting an acid because heat is generated during dilution

- Make sure to turn gas lines to Bunsen burners off at the completion of a task.

ACCURACY

Accuracy is also important when performing a laboratory task. Use the following guidelines to ensure accuracy of your experimental results:

- Mix liquid chemicals slowly and thoroughly to ensure homogeneity.

- Do not contaminate chemicals by inserting a utensil into a bottle. Pour an approximate quantity of chemical into another container and then aliquot appropriately. Never return excess chemicals to the bottle.

- When weighing chemicals, do not weigh directly on the scale. Use a separate container to avoid contamination and prevent corrosion of the scale.

- When weighing glassware, allow hot glassware to cool to room temperature. Hot objects experience convection currents and will appear to weigh less than they actually do.

- Pay attention to significant figures when recording experimental results.

SIGNIFICANT FIGURES

Keep in mind that your experimental results can only be as precise as the measurements you took, and their precision is determined by the precision of the measurement tools you used. The more significant figures, the more precise the results.

Use the following guidelines when recording results, using the appropriate amount of significant figures:

- All zeros between digits and nonzero digits are significant.

    987 has three significant figures

    3.565 has four significant figures

    45.793 has five significant figures

- Zeros to the left of the first nonzero digits are not significant.

    0.0006 has 1 significant figure

    0.00703 has 2 significant figures

- Zeros to the right of the last nonzero digit are significant.

  7.00 has 3 significant figures

  0.08670 has 4 significant figures

LABORATORY PROCEDURES

Separation Methods

Filtration
Filtration is a method used to separate solids from liquids. Filters are generally made from porous paper materials that will trap solids and allow liquid to pass through. In order to accurately determine how much solid is present, the filter paper is weighed when it's dry, prior to the filtration process, and weighed again once the resulting solid and the paper are allowed to dry. The difference in the two weights is the weight of the solid sample.

Chromatography
Chromatography is a method used to separate substances based on their specific properties. Paper chromatography is a separation method that measures the migration of compounds along a strip of paper, with solvents representing the mobile phase. Substances that have a greater tendency to stick to the paper surface will migrate more slowly than substances that are less attracted to the paper surface.

Titration
Titration methods are used to determine the concentration of an acid or a base. Since it takes one mole of hydrogen ions to neutralize one mole of hydroxide ions (and vice versa), a known concentration of acid or base can be used to titrate into an unknown concentration solution. The unknown concentration can be determined by the quantity of known solution that was used to neutralize it.

$$\text{Molarity} = \frac{\text{moles}}{\text{Liter}}$$

$$\text{moles} = \text{Molarity} \times \text{Liters}$$

An indicator such as litmus paper or phenolphthalein can be used to determine the equivalence point when the sample has been neutralized. Phenolphthalein will be clear in acidic solutions (pH < 7) and hot pink in basic solutions (pH > 7). Litmus paper is red in acidic solutions and blue in basic solutions (remember Blue = Basic).

Identifying Chemicals

Precipitation
Ions in solution can be identified by forcing precipitation. If you know whether a salt is soluble or insoluble, you may be able to deduce which ions are present in aqueous solutions. For example, most chlorine salts are soluble in water, but silver chloride is not. If you were to add chloride ions into an aqueous solution and a precipitate formed, then your solution most likely contains silver ions.

Flame Tests
Flame tests are used to identify alkali metals (Li^+, Na^+, K^+,...) and alkaline earth metals (Ba^{2+}, Sr^{2+}, and Ca^{2+},...), which will both produce colored flames when burnt.

Colored Solutions
Many of the transition metals dissociate into ions in solution and will form colored solutions. For example, permanganate (MnO_4^-), a strong oxidizing agent, will turn purple in solution and dichromate ($C_2O_7^-$), another strong oxidizing agent, will turn orange in solution.

Conduction
Ions in solution will conduct electricity and nonionic solutions will not. If you wanted to determine whether or not a solution contained ions, you could measure the solution's conductivity.

SAMPLE LABORATORY ASSIGNMENT

Laboratory Materials

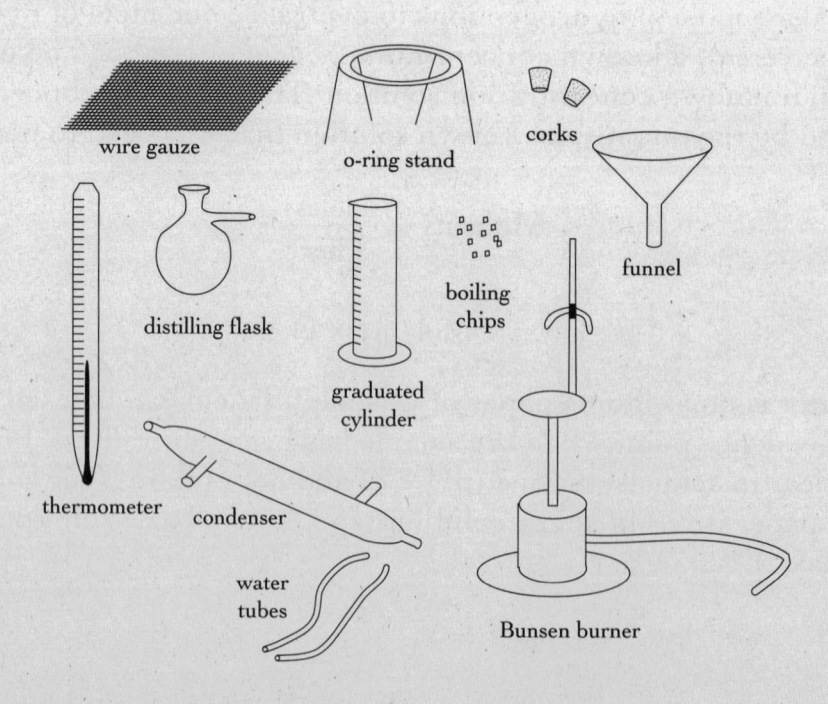

Distillation

Distillation is a method used to identify a pure liquid substance. A simple distillation apparatus will be set up to perform this method in an unknown pure substance. The temperature and volume of the distillate will be obtained and recorded. The resulting data will be graphed in order to determine the boiling point of the unknown substance. This data will be compared to a reference table in order to identify and report the unknown substance.

The distillation method has been performed for thousands of years as a purification tool. Due to the increased efficiency demands for this process, initiated by the beverage industry, the distillation process was updated and finally optimized by the 1920s. The process involves converting a liquid to a gas through boiling and then condensing the gas into a liquid.

The simple distillation apparatus is comprised of three parts.

- A distillation flask in which the initial substance is boiled.

- A distillation tube or condenser in which gas will be condensed back into a liquid.

- A container (usually a graduated cylinder) to receive the distillate from the condenser.

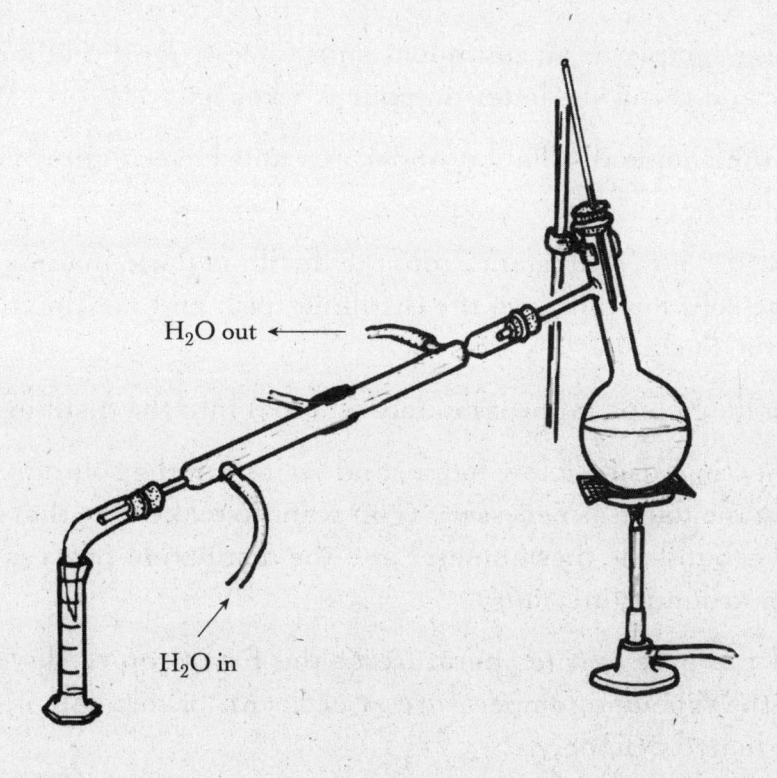

LABORATORY ♦ 149

A pure liquid will be distilled completely when the substance is brought to a certain temperature.

When performing this experiment, keep in mind the following:

Use an appropriately sized distillation flask (one that will be approximately half-full at the beginning of the experiment).

Use the appropriately sized thermometer—see your laboratory instructor.

Use corks between the distillation flask and thermometer, and between the distillation flask and the condenser in order to avoid loss of vapor.

Make sure the thermometer is placed low enough in the distillation flask to allow vapor to surround the bulb when the experiment commences.

Make sure the distillation flask and the condenser are clamped to a fixed pole on the laboratory bench.

Be sure to add boiling chips to the distilling flask to avoid boiling the substance too quickly.

Use a wire gauze placed on a stand between the flame and distilling flask to prevent the flame from hitting the glass surface of the flask.

Keep a constant circulation of cool water through the jacket of the condenser. The cold water should be added to the bottom of the tube and released at the top end.

Procedure

1. Receive a sample of an unknown substance from the laboratory instructor and record its letter on your worksheet.

2. Set up the simple distillation apparatus and have it approved before proceeding.

3. Pour the unknown substance into the distilling flask, making sure that all of the solution falls into the distilling flask and not down the side-arm of the flask.

4. Place boiling chips (approximately 6 chips) into the distilling flask.

5. Begin heating, using a low flame, and wait until the solution is boiling to adjust the flame as necessary (you want to make sure that a vapor is created around the thermometer and the distillation process occurs at a rate of around 2 mL/min.)

6. Record the time and temperature at the first drop of distillate, and record the time and temperature at each mL of distillate collected in the graduated cylinder.

7. After the distillation process is completed, discard the residuals and distillate in the appropriate containers.

Calculation

1. Graph the distillation temperature verses the total volume of the distillate.

$$y\text{-axis} = \text{temperature}$$

$$x\text{-axis} = \text{volume}$$

2. Determine the boiling point of the unknown substance and record the information.

3. Compare the results to the following table in order to determine the identity of the substance.

Total volume of distillate, ml	Distillation temperature, °C	Time, min	Total volume of distillate, ml	Distillation temperature, °C	Time, min
First drop			13.0		
1.0			14.0		
2.0			15.0		
3.0			16.0		
4.0			17.0		
5.0			18.0		
6.0			19.0		
7.0			20.0		
8.0			21.0		
9.0			22.0		
10.0			23.0		
11.0			24.0		
12.0			25.0		

LABORATORY ◆ 151

Chapter 16

PRACTICE TEST ONE

1. A + B + C → D + E

Experiment	[A]	[B]	[C]	Reaction rate (mol/L × sec)
1	0.1	0.1	0.1	2.0×10^{-3}
2	0.2	0.1	0.1	4.0×10^{-3}
3	0.1	0.1	0.2	4.0×10^{-3}
4	0.1	0.2	0.1	2.0×10^{-3}

According to the experimental data collected and reported above, which of the following would cause an increase in the reaction rate?

I. Increasing [A]

II. Increasing [B]

III. Increasing [C]

A. I only

B. I and II only

C. I and III only

D. I, II, and III

2. Which of the following is correct concerning lithium and fluorine?

A. Fluorine has greater electronegativity and larger first ionization energy.

B. Fluorine has larger atomic radius and greater electronegativity.

C. Lithium has greater electronegativity and larger first ionization energy.

D. Lithium has a greater atomic radius and a larger first ionization energy.

3. A 5 L container holds an ideal gas at 5 atm. If the gas is then transferred to a new tank of 15 L and temperature is kept constant, what would the new pressure be?

A. $\frac{3}{5}$ atm

B. $\frac{5}{3}$ atm

C. 5 atm

D. 15 atm

4. How many calories of heat are required to bring a 1 g sample of water from –25 °C to 50 °C? (Heat of fusion for water = 80 cal/g, specific heat of ice = 0.5 cal/g °C, specific heat of liquid water = 1 cal/g °C.)

A. 62.5

B. 80

C. 142.5

D. 155

5. Which of the following orbital hybridizations is able to form molecules that are trigonal planar or bent?

A. *sp*

B. *sp²*

C. *sp³*

D. *dsp²*

6. What is the freezing point of a 5 M solution of potassium bromide in water at standard pressure? (K_f = 1.9 °C/m)

   A. −19.0 °C
   B. −9.5 °C
   C. 0 °C
   D. 19.0 °C

7. Which of the following molecules is polyprotic?

   A. HCl
   B. HClO$_4$
   C. HNO$_3$
   D. H$_2$SO$_4$

8. Which of the following is true regarding CH$_4$?

   I. Molecular shape is tetrahedral
   II. Is a polar molecule
   III. Will experience hydrogen bonding

   A. I only
   B. II only
   C. II and III only
   D. I, II, and III

9.

Bond	Bond Energy
H — H	436 kJ
O = O	499 kJ
O — H	463 kJ

Using the bond energies above, what is the ΔH for the following reaction?

$$2H_2(g) + O_2(g) \rightarrow 2H_2O(g)$$

   A. −908 kJ
   B. −481 kJ
   C. 481 kJ
   D. 908 kJ

10. If a compound contains 75% mercury and 25% chlorine, what is the empirical formula?

    A. HgCl
    B. Hg$_2$Cl
    C. HgCl$_2$
    D. HgCl$_3$

11. According to the following equation, 16 g of O$_2$ reacted with Cu$_2$S will produce how many grams of Cu?

    $$3\,Cu_2S + 3\,O_2 \rightarrow 3\,SO_2 + 6\,Cu$$

    A. 15.9 g
    B. 31.75 g
    C. 63.5 g
    D. 127.0 g

12. A sealed container holds a mixture of nitrogen, oxygen, and other gases. Nitrogen has a pressure of 5 atm, oxygen has a pressure of 2 atm, and the other gases have a pressure of 3 atm. What percentage of the partial pressure is oxygen?

    A. 20%
    B. 30%
    C. 50%
    D. 70%

13. If it takes 80.0 mL of 0.5 M HCl reacting with an unknown quantity of NaOH to complete the following reaction, how many grams of NaOH must have reacted?

    $$NaOH(aq) + HCl(aq) \rightarrow NaCl(aq) + H_2O(l)$$

    A. 0.016 g
    B. 1.6 g
    C. 160 g
    D. 1600 g

14. A sample of an unknown gas weighs 10 g and occupies 5.6 L at standard temperature and pressure. Which of the following is most likely the gas?

    A. H_2
    B. He
    C. Ar
    D. Ne

15. When pressure is increased and temperature is held constant, which of the following phase changes can occur?

    I. Freezing
    II. Condensation
    III. Deposition

    A. I only
    B. I and II only
    C. II only
    D. I, II, and III

16. Which of the following Lewis dot structures correctly represents O_2?

    A. :O=O:
    B. ::O=O::
    C. :::O–O:::
    D. :::O=O:::

17. How many grams of sodium acetate ($NaC_2H_3O_2$) would be needed to create a 500 g aqueous solution that contains 6.0% sodium acetate?

    A. 6 g
    B. 30 g
    C. 300 g
    D. 470 g

18. What can we determine, knowing that we have a 1.0 m $CaCl_2$ aqueous solution?

   I. The mole fraction of $CaCl_2$ in the solution

   II. The total number of grams of $CaCl_2$

   III. The number of grams of solvent

   A. I only
   B. I and II only
   C. II and III only
   D. I, II, and III

19. Consider the following reaction:

   $$2HI(g) + Cl_2(g) \leftrightarrow 2HCl(g) + I_2(g) + energy$$

   Which of the following will cause an increase in the number of I_2 moles?

   A. An increase in volume while holding temperature constant
   B. A decrease in volume while holding temperature constant
   C. An increase in temperature while holding volume constant
   D. A decrease in temperature while holding volume constant

20. What is the K_{sp} of copper(I) bromide if the solubility is 2×10^{-4} mol/L at 25 °C?

   A. 4×10^{-16}
   B. 4×10^{-8}
   C. 2×10^{-8}
   D. 2×10^{-2}

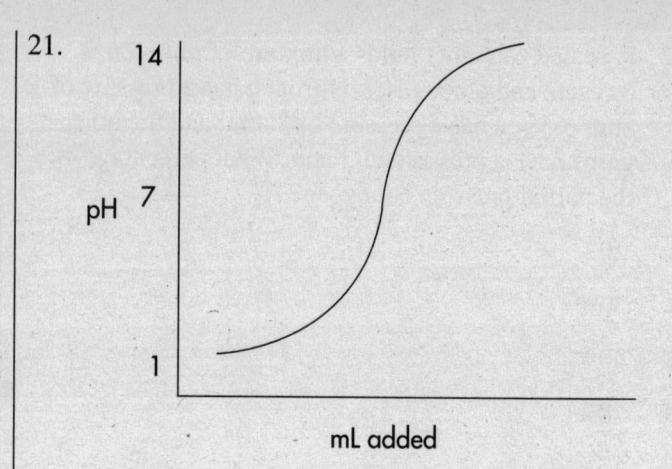

21.

The diagram above could represent which of the following?

   A. The addition of NaOH to HCl
   B. The addition of HCl to NaOH
   C. The addition of HF to KOH
   D. The addition of H_2SO_4 to $Ba(OH)_2$

22. If a reaction is spontaneous, which of the following must be correct?

   A. The reaction is exothermic
   B. The reaction is endothermic
   C. ΔS is negative
   D. ΔG is negative

23.

$$A + B + 2C \rightarrow 2D + 2E$$

The rate law for the equation above is:

Rate = $k[A][B]$

Which of the following will increase the reaction rate?

A. Decreasing [A]
B. Decreasing [B]
C. Increasing [B]
D. Increasing [C]

24. The electron configuration for a chloride ion in its ground state is most likely represented by which of the following?

A. $1s^2\ 2s^2\ 2p^6\ 3s^2$
B. $1s^2\ 2s^2\ 2p^6\ 3s^2\ 3p^5$
C. $1s^2\ 2s^2\ 2p^6\ 3s^2\ 3p^6$
D. $1s^2\ 2s^2\ 2p^6\ 3s^2\ 3p^6\ 4s^2$

25. Which of the following order of elements represents increasing atomic radius?

A. Na, Mg, Si, Cl
B. Cs, Rb, Na, Li
C. Li, B, O, Ne
D. Be, Mg, Ca, Sr

26. A sample of gas contains 0.6 mol of hydrogen and 1.4 mol of argon. If the sample is held at STP, what is the partial pressure resulting from the argon gas?

A. 0.6 atm
B. 0.7 atm
C. 1.0 atm
D. 1.4 atm

27. What is the percent composition by mass of the elements in $CaCO_3$?

A. 20% Ca, 20% C, 60% O
B. 40% Ca, 12% C, 48% O
C. 20% Ca, 24% C, 56% O
D. 40% Ca, 12% C, 32% H

28. According to Graham's law, which of the following gases would most likely have a rate of effusion that is four times the rate of H_2?

A. O_2
B. N_2
C. CO_2
D. H_2O

29. Which of the following statements regarding the order in which electrons fill orbital is NOT true?

A. Closest to the nucleus to farthest from the nucleus
B. Lowest energy to highest energy
C. Greatest stability to least stability
D. Largest orbital to smallest orbital

30. The following reaction represents the production of nitrogen gas from combustion of ammonia in oxygen.

$$4NH_3(g) + 3O_2(g) \rightarrow 2N_2(g) + 6H_2O(g)$$
$$\Delta H = -1270 \text{ kJ}$$

If 1867 kJ of energy is released, how many grams of ammonia were present at the start of the reaction?

A. 1 g

B. 10 g

C. 100 g

D. 1000 g

31. Which of the following nuclides is a result of $^{241}_{84}Po$ after it emits 2 alpha particles and 1 beta particle?

A. $^{206}_{82}Pb$

B. $^{233}_{81}Ti$

C. $^{207}_{80}Hg$

D. $^{207}_{81}Ti$

32. If a sample of propane (C_3H_8) reacted completely with O_2 at 760 mmHg and 25 °C to yield 720 mL of CO_2 gas, how many grams of C_3H_8 were consumed in the reaction?

$$C_3H_8(g) + O_2(g) \rightarrow 3CO_2(g) + 4H_2O(g)$$

A. 0.44 g

B. 3.96 g

C. 440 g

D. 3960 g

WRITTEN RESPONSE QUESTION

Many reactions that occur in aqueous environments are a result of ions dissociating into solution and reforming bonds. For example, calcium carbonate is formed by reacting calcium hydroxide with sodium carbonate in an aqueous solution. Calcium carbonate is the main ingredient in many antacid products; it is a neutralizing agent.

Describe how ions play a role in the process of producing calcium carbonate in an aqueous solution.

Explain how calcium carbonate can be used as neutralizing agent.

Use diagrams and chemical reactions in your answer choices.

The Princeton Review

YOUR NAME: _____
(Print)   Last   First   M.I.

SIGNATURE: _____ DATE: __/__/__

HOME ADDRESS: _____
(Print)   Number and Street

_____
         City   State   Zip Code

PHONE NO.: _____
(Print)

Completely darken bubbles with a No. 2 pencil. If you make a mistake, be sure to erase mark completely. Erase all stray marks.

Practice Test One

1. Ⓐ Ⓑ Ⓒ Ⓓ
2. Ⓕ Ⓖ Ⓗ Ⓙ
3. Ⓐ Ⓑ Ⓒ Ⓓ
4. Ⓕ Ⓖ Ⓗ Ⓙ
5. Ⓐ Ⓑ Ⓒ Ⓓ
6. Ⓕ Ⓖ Ⓗ Ⓙ
7. Ⓐ Ⓑ Ⓒ Ⓓ
8. Ⓕ Ⓖ Ⓗ Ⓙ
9. Ⓐ Ⓑ Ⓒ Ⓓ
10. Ⓕ Ⓖ Ⓗ Ⓙ
11. Ⓐ Ⓑ Ⓒ Ⓓ
12. Ⓕ Ⓖ Ⓗ Ⓙ
13. Ⓐ Ⓑ Ⓒ Ⓓ
14. Ⓕ Ⓖ Ⓗ Ⓙ
15. Ⓐ Ⓑ Ⓒ Ⓓ
16. Ⓕ Ⓖ Ⓗ Ⓙ

17. Ⓐ Ⓑ Ⓒ Ⓓ
18. Ⓕ Ⓖ Ⓗ Ⓙ
19. Ⓐ Ⓑ Ⓒ Ⓓ
20. Ⓕ Ⓖ Ⓗ Ⓙ
21. Ⓐ Ⓑ Ⓒ Ⓓ
22. Ⓕ Ⓖ Ⓗ Ⓙ
23. Ⓐ Ⓑ Ⓒ Ⓓ
24. Ⓕ Ⓖ Ⓗ Ⓙ
25. Ⓐ Ⓑ Ⓒ Ⓓ
26. Ⓕ Ⓖ Ⓗ Ⓙ
27. Ⓐ Ⓑ Ⓒ Ⓓ
28. Ⓕ Ⓖ Ⓗ Ⓙ
29. Ⓐ Ⓑ Ⓒ Ⓓ
30. Ⓕ Ⓖ Ⓗ Ⓙ
31. Ⓐ Ⓑ Ⓒ Ⓓ
32. Ⓕ Ⓖ Ⓗ Ⓙ

Multiple-Choice Answer Key

1. C
2. A
3. B
4. C
5. B
6. A
7. D
8. A
9. B
10. C
11. C
12. A
13. B
14. C
15. D
16. B
17. B
18. A
19. D
20. B
21. A
22. D
23. C
24. C
25. D
26. B
27. B
28. A
29. D
30. C
31. B
32. A

Practice Test One Answers and Explanations

1. **C.**

    You must first determine the rate law. Comparing experiments 1 and 2, when the [B] and [C] are held constant and [A] is doubled, the rate of reaction is also doubled. The reaction is therefore first order with respect to [A]. Comparing experiments 1 and 3, when [A] and [B] are held constant and [C] is doubled, the reaction also doubles. The reaction is first order with respect to [C]. Comparing experiments 1 and 4, [A] and [C] are held constant and [B] is doubled, the reaction rate stays the same. The reaction is then zero order with respect to [B]. Therefore, the rate law is: $= k[A][C]$.

    Only by increasing [A] and [C] will the reaction rate increase, since the reaction is only dependent on [A] and [C].

2. **A.**

    Generally speaking, as we move across a horizontal row of the periodic table (from left to right), electronegativity and ionization energy increases while atomic radius decreases. As we move across the periodic table, protons are added to the nucleus without adding shells. More protons will exert a stronger pull on the outer electrons, resulting in a decrease in radius and an increase in electronegativity and ionization energy.

3. **B.**

    Pressure and volume are inversely proportional. If you increase volume, pressure decreases. Using Process of Elimination, you can cross off choices C and D since the pressure value is greater or equal to the original.

    $$P_1 V_1 = P_2 V_2$$

    $$P_2 = \frac{P_1 V_1}{V_2} = \frac{(5)(5)}{(15)} = \frac{25}{15} = \frac{5}{3} \text{ atm}$$

4. **C.**

    Calculate in steps:
    $$-25 \, °C \rightarrow 0 \, °C$$
    $$q = mc\Delta T = (1)(0.5)(25) = 12.5 \text{ cal}$$

    Phase change: solid $\rightarrow$ liquid = 80 cal
    $$0 \, °C \rightarrow 50 \, °C$$
    $$q = mc\Delta T = (1)(1)(50) = 50 \text{ cal}$$

    Total calories needed =
    12.5 + 80 + 50 = 142.5 cal

5. **B.**

    The sp^2 hybrid orbital forms bent molecules (i.e., SO_2) when the central atom has two bonds and one unshared pair of electrons. These orbitals form trigonal planar molecules (i.e., BF_3) when the central atom is bonded to three atoms and has no unshared pair of electrons.

6. **A.**

   The freezing point of water will be decreased from 0 °C. Therefore, cross off answer choices C and D since they are not below zero.

   $\Delta T = k_f mi = (1.9)(5)(2) = 19$ °C

   The freezing point of water is 0 °C. The freezing point of the aqueous solution is then:

   $0 - 19 = -19$ °C

7. **D.**

   Polyprotic means that the molecule has more than one hydrogen ion to give up. Choice D is the only molecule listed that can donate more than one hydrogen ion. Since it has two, it can be further classified as diprotic.

8. **A.**

   The central carbon atom has four electron pairs, which tend to yield a tetrahedral shape. Choice B is not correct since there is no net dipole moment in a molecule of CH_4. Since CH_4 does not have a net dipole moment, it is a nonpolar molecule. Therefore, choices C and D (which both mention item III) are incorrect. Helpful hint: Once you have determined A to be correct, you can automatically cross off choices B and D since they do not include item I.

9. **B.**

   ΔH = Energy of the bonds broken − energy of the bonds formed

   $= 2(H) + 1(O) - 4(OH)$

   $= 2(436) + 1(499) - 4(463)$

   Approximate:

   $= 2 \times 450 + 500 - 1850 = -450$ kJ

   (which is closest to B)

10. **C.**

    First calculate the quantity in a 100 g sample from the percentages. (Remember to use approximations when doing calculations.)

    75% Hg = 75 g Hg

    25% Cl = 25 g Cl

    Next convert grams to moles:

    $70 \text{ g Hg} \times \dfrac{1 \text{ mol}}{200 \text{ g}} = 0.375$ mol Hg

    $25 \text{ g Cl} \times \dfrac{1 \text{ mol}}{35.5 \text{ g}} = 0.7$ mol Cl

    Next calculate the mole-to-mole ratio between the elements:

    $\dfrac{\text{Hg}}{\text{Cl}} = \dfrac{0.375 \text{ mol Hg}}{0.7 \text{ mol Cl}} = \dfrac{1 \text{ mol Hg}}{2 \text{ mol Cl}}$

    The empirical formula is therefore $HgCl_2$.

11. **C.**

First convert grams of O_2 into moles.

$$16 \text{ g } O_2 \times \frac{1 \text{ mol}}{32 \text{ g}} = 0.5 \text{ mol } O_2$$

According to the balanced equation, every 3 moles of O_2 react to produce 6 moles of Cu. Therefore,

$$0.5 \text{ mol } O_2 \times \frac{6 \text{ mol Cu}}{3 \text{ mol } O_2} = 1 \text{ mol Cu}$$

Finally convert moles of Cu into grams:

$$1 \text{ mol Cu} \times \frac{63.5 \text{ g}}{1 \text{ mol}} = 63.5 \text{ g Cu}$$

12. **A.**

Total pressure =
$P_{gas1} + P_{gas2} + P_{gas3} = 5 + 2 + 3 = 10$ atm

The oxygen gas makes up a pressure of 2 atm out of the total pressure of 10 atm.

$$\frac{2}{10} = 20\%$$

13. **B.**

First convert the 80 mL of 0.5 M HCl into moles:

$$M = \frac{\text{mol}}{\text{L}}$$

$$\text{mol} = M \times L = 0.5M \times 0.080 \text{ L} = 0.04 \text{ mol HCl}$$

According to the balanced equation, every mole of HCl reacts with 1 mole of NaOH. Therefore:

0.04 mol HCl = 0.04 mol NaOH

Finally, convert moles of NaOH into grams. You need to calculate the molecular weight of NaOH first.

MW = (23) + (16) + (1) = 40 g/mol

$$0.04 \text{ mol NaOH} \times \frac{40 \text{ g}}{1 \text{ mol}} = 1.6 \text{ g NaOH}$$

14. **C.**

Since 5.6 is $\frac{1}{4}$ the standard volume (22.4L), the number of moles is equal to $\frac{1}{4}$, or 0.25.

$$\frac{10 \text{ g}}{.25 \text{ mol}} = 40 \text{ g / mol}.$$

The molecular weight of Ar is 40 g/mol.

15. **D.**

All three phase changes can occur as pressure is increased and temperature is held constant as shown in the diagram below.

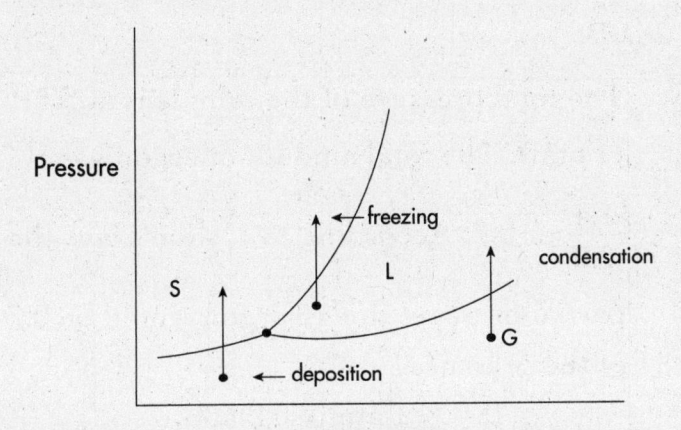

16. **B.**

The total number of valence electrons in O_2 is 12. [$2 \times 6e^- (O) = 12 \; e^-$] The only answer choice with 12 electrons is B.

17. **B.**

Mass of $NaC_2H_3O_2 = 0.06 \times 500 \; g = 30$ grams

18. **A.**

We can determine the mole fraction since we know the molality of the solution. Molality is moles of solute per kilogram of solvent. Once we convert kilograms of solvent to moles, we can determine the mole fraction for either $CaCl_2$ or the water. Therefore, item I is correct, and we can cross off choice C, since it does not contain item I.

We are unable to determine the total grams of solute or solvent since we do not know the total amount of solution.

19. **D.**

According to Le Chatelier's principle, a disruption in equilibrium will cause a shift in the direction of the reaction. By decreasing temperature, the reaction will proceed in the direction that produces energy. Thus, the forward reaction will be favored and additional I_2 moles will be produced.

Changing the volume (choices A and B) in this example will not affect the equilibrium, since there are an equal number of moles on both sides of the equation.

20. **B.**

The reaction is as follows:

$CuBr(s) \rightarrow \quad Cu^+(aq) + \quad Br^-(aq)$

$ \; 2 \times 10^{-4} \quad\quad 2 \times 10^{-4} \quad\quad 2 \times 10^{-4}$

$K_{sp} = [Cu^+][Br^-] = [2 \times 10^{-4}][2 \times 10^{-4}] = 4 \times 10^{-8}$

21. **A.**

The diagram represents the titration of an acid by a base. The pH starts off at around 1 (acidic) and rises to 13 (basic). Choices B, C, and D represent the titration of a base by an acid.

22. **D.**

The change in free energy is represented by ΔG.

$$\Delta G = \Delta H - T\Delta S$$

All reactions that experience a decrease in free energy are spontaneous ($\Delta G < 0$). Endothermic reactions have a positive ΔH, meaning that energy is absorbed. An endothermic reaction is generally not spontaneous. Although energy is released with an exothermic reaction, if the ΔS value is large enough, an exothermic reaction CAN be nonspontaneous. Remember, the question asks for which statement MUST be true. A negative or positive ΔS can be spontaneous or nonspontaneous, depending on the ΔH value.

23. **C.**

Since the reaction is not dependent on [C], the reaction rate will only be affected by increasing either [A] or [B].

24. **C.**

Chlorine forms a bond by accepting an electron from its bonding partner. It becomes a negatively charged ion with the same electron configuration as argon. Watch out: neutral chlorine in its ground state is represented by choice B.

25. **D.**

Generally speaking, atomic radius increases as you move down a group (vertical column) and decreases as you move across a period (horizontal row).

26. **B.**

The total pressure of the sample is at STP (1 atm). The total amount of argon gas is $\frac{1.4}{2.0}$ or 0.7. According to Dalton's law, the partial pressure due to argon would be 0.7 of the pressure.

$$0.7 \times 1 \text{ atm} = 0.7 \text{ atm}.$$

27. **B.**

You can automatically cross off choice D because the total percentage does not equal 100%.

Calculate the molecular weight of $CaCO_3$.

$$MW = 40 + 12 + 3(16) = 100 \text{ g/mol}$$

The percent by mass of Ca is $\frac{40}{100} = 40\%$

The percent by mass of C is $\frac{12}{100} = 12\%$

The percent by mass of O is $\frac{48}{100} = 48\%$

Watch out for choice A. Remember, the question asks for percent composition by mass.

28. A.

According to Graham's law:

$$\frac{\frac{v_1}{v_2}}{\sqrt{\frac{m_1}{m_2}}}$$

According to this equation, in order to obtain a rate of effusion that is four times that of H_2, the mass or molecular weight of the molecule must be 16 times as large.

The molecular weight of H_2 is 2; therefore, the molecule that weighs 16 × 2 (or 32) is O_2.

29. D.

Generally speaking, orbitals are filled from closest to the nucleus to the farthest. The smaller orbitals are generally filled first since they are closest to the nucleus.

30. C.

If 1270 kJ of energy are released per 4 moles of NH_3, then:

$$1867 \text{ kJ} \times 4 \text{ mol } NH_3 \approx 6 \text{ mol } NH_3$$

$$6 \text{ mol } NH_3 \times \frac{17 \text{ g } NH_3}{1 \text{ mol } NH_3} \approx 100 \text{ g}$$

31. B.

If the original nuclide emits two alpha particles, which is like emitting two helium ions $^{4}_{2}He$, the mass number will be reduced by four for each alpha particle and the atomic number will be reduced by one for each particle. The beta particle, which is equivalent to an electron, will result in no change in the mass number and an increase in the atomic number by one.

$$^{241}_{84}Po - ^{4}_{2}\alpha - ^{4}_{2}\alpha - ^{0}_{-1}\beta = ^{233}_{81}Ti$$

32. A.

First convert all terms into the appropriate units for the ideal gas equation:

Pressure: 760 mmHg = 1 atm

Temperature: 25 °C = 298K ≈ 300K

Volume: 720 mL = 0.720 L

$PV = nRT$

$$n = \frac{PV}{RT} = (1)(0.720)/(0.08)(300) = 0.03$$
moles CO_2

According to the balanced equation, 1 mole of propane will produce 3 moles of carbon dioxide. Therefore:

$$0.03 \text{ moles } CO_2 \times \frac{1 \text{ mol } C_3H_8}{3 \text{ mol } CO_2} = 0.01 \text{ mol } C_3H_8$$

Finally, convert moles of propane into grams. You need to calculate molecular weight first.

MW = 3(12) + 8(1) = 44 g/mol

0.1 mol C_3H_8 × 44 g/1 mol = 0.44 g

WRITTEN RESPONSE QUESTION

One way to produce calcium carbonate is by the reaction of calcium hydroxide and sodium carbonate in an aqueous solution.

$$Ca(OH)_2 \ (aq) + Na_2CO_3 (aq) \rightarrow CaCO_3 \ (s) + 2NaOH \ (aq)$$

This reaction involves ions in solution. Solid calcium hydroxide dissolves in water to produce Ca^{2+} (aq) and $2OH^-$(aq) ions, and solid Na_2CO_3 dissolves in water to produce $2Na^+$ (aq) and CO_3^{2-} ions. The following ionic equation illustrates this reaction:

$$Ca^{2+} \ (aq) + 2OH^- \ (aq) + 2Na^+(aq) + CO_3^{2-} \ (aq) \rightarrow CaCO_3 \ (s) + 2Na^+(aq) + 2OH^-(aq)$$

Calcium carbonate can be used to neutralize acid solutions. Carbonates that react with acids yield gas products.

$$CaCO_3 \ (aq) + 2HCl \ (aq) \rightarrow CaCl_2 \ (s) + H_2O(l) + CO_2 \ (g)$$

Chapter 17

PRACTICE TEST TWO

1. Which of the following is a diamagnetic element?

   A. Oxygen

   B. Phosphorous

   C. Chlorine

   D. Calcium

2. Which of the following represents the correct noble gas configuration for rubidium (Rb)?

   A. [Kr] $4s^1$

   B. [Kr] $5s^1$

   C. [Ar] $4s^1$

   D. [Ar] $5s^1$

3. Which of the following is correct regarding the following reaction?

   $N_2(g) + 2H_2(g) \rightarrow N_2H_4(l)$

   $N_2H_4(l) + O_2(g) \rightarrow N_2(g) + 2H_2O(l)$ $\quad \Delta H = -622$ kJ

   $H_2(g) + \frac{1}{2}O_2(g) \rightarrow H_2O(l)$ $\quad \Delta H = -286$ kJ

   A. $\Delta H = 50$ kJ and the reaction is exothermic

   B. $\Delta H = 50$ kJ and the reaction is endothermic

   C. $\Delta H = 336$ kJ and the reaction is exothermic

   D. $\Delta H = 336$ kJ and the reaction is endothermic

4. Which of the following orbital hybridizations is able to form molecules that are linear?

   A. sp

   B. sp^2

   C. sp^3

   D. dsp^2

5. $2SO_2(g) + O_2(g) \leftrightarrow 2SO_3(g) +$ energy

   Which of the following would cause the equilibrium expression above to shift to the left?

   A. Decreasing temperature

   B. Increasing $[O_2]$

   C. Increasing volume

   D. Decreasing $[SO_3]$

6. If a 9.2 g sample contains only 2.8 g of nitrogen and 6.4 g of oxygen, what is the empirical formula of the compound?

   A. NO

   B. N_2O

   C. NO_2

   D. NO_3

PRACTICE TEST TWO ◆ 177

7.

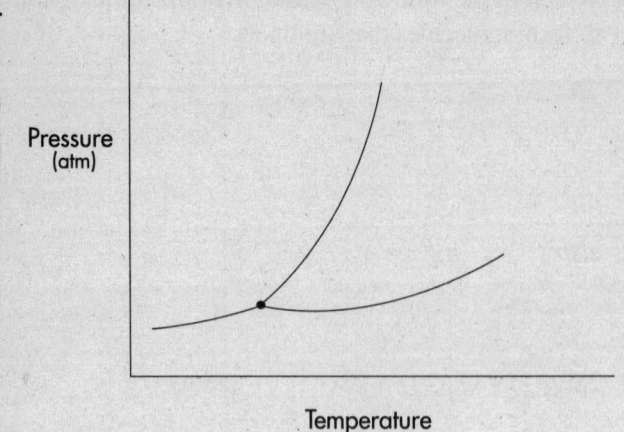

Pressure (atm) vs Temperature phase diagram

Which of the following phase changes could occur as a result of pressure being decreased and temperature being increased?

A. Liquid to solid to gas
B. Solid to liquid to gas
C. Liquid to gas to solid
D. Gas to liquid to solid

8. Which of the following Lewis dot structures correctly represents Cl_2?

A. :::Cl-Cl:::
B. ::Cl=Cl::
C. ::Cl=Cl⁻::
D. :::Cl=Cl:::

9. Atmospheric nitrogen (N_2) occupies a container of 90 mL at a temperature of 50 °C and a pressure of 2280 mm Hg. How many grams of the gas are present? (R = 0.0821 L • atm/mol • K)

A. 0.47 g
B. 12.5 g
C. 175 g
D. .3 g

10. Which of the following is true regarding H_2O.

  I. Molecular shape is linear
  II. Will experience hydrogen bonding
  III. Orbital hybridization is sp^3

A. I only
B. II only
C. II and III only
D. I, II, and III

11. What is the mass of oxygen in 170 g of $NaNO_3$?

A. 16 g
B. 48 g
C. 64 g
D. 96 g

12. The free metal iron (Fe) can be extracted from the iron ore hematite (Fe_2O_3), by reacting the ore with carbon monoxide (CO). This reaction is as follows:

$$Fe_2O_3(s) + 3CO(g) \rightarrow 2Fe(s) + 3CO_2(g)$$

If 32.3 g of iron ore is reacted with carbon monoxide, how many grams of iron can be extracted?

A. 5.6 g
B. 22.6 g
C. 44.8 g
D. 57.3 g

13. What is the mass of hydrogen in 540 g of acetic acid ($HC_2H_3O_2$)?

    A. 9 g
    B. 36 g
    C. 216 g
    D. 288 g

14. Eight grams of helium gas are sealed in a 2L container at a pressure of 2 atm. At constant temperature, if the volume of the gas was increased to 8 L, what would the pressure be?

    A. 0.5 atm
    B. 2 atm
    C. 4 atm
    D. 8 atm

15. If $^{210}_{82}Pb$ emits one alpha and two beta particles, what is the resulting nuclide?

    A. $^{206}_{80}Hg$
    B. $^{206}_{82}Pb$
    C. $^{214}_{83}Bi$
    D. $^{213}_{83}Bi$

16. According to Graham's law, which of the following gases would most likely have the lowest rate of effusion within the same environment?

    A. H_2
    B. O_2
    C. N_2
    D. He

17. In order to increase the temperature of 1 g of water from a $-10\,°C$ to $110\,°C$, how many calories of heat are needed? (Heat of fusion for water = 80 cal/g, heat of vaporization for water = 540 cal/g, specific heat of ice and steam = 0.5 cal/g °C, specific heat of liquid water = 1 cal/g °C)

    A. 110
    B. 270
    C. 620
    D. 730

18. Which point on the following diagram represents the boiling point of a substance?

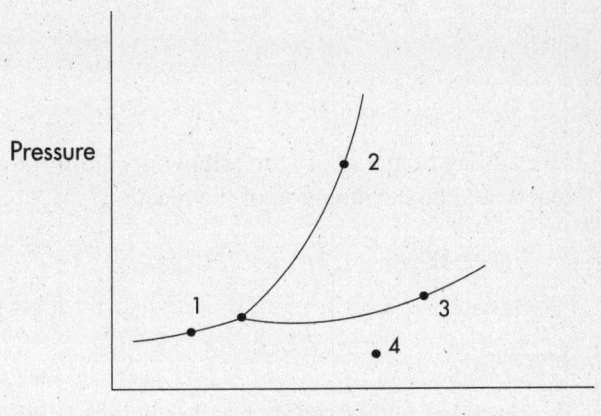

    A. 1
    B. 2
    C. 3
    D. 4

19. What is the boiling point of a 3 m aqueous solution of CaCl$_2$ at 1 atm? (k_b = 0.52 °C/m)

A. 100.0 °C

B. 101.5 °C

C. 103.0 °C

D. 104.5 °C

20. Calculate the molality of an aqueous solution that contains 9 g of glucose, C$_6$H$_{12}$O$_6$, dissolved in 20 g of water.

A. 0.0025 m

B. 0.025 m

C. 0.25 m

D. 2.5 m

21. If we add hydrogen gas to the following equilibrium, what would be the direction of the reaction?

$$H_2(g) + I_2(g) \leftrightarrow 2HI(g)$$

A. Forward

B. Reverse

C. The reaction will not shift but will remain at equilibrium

D. The reaction would stop

22. After adding NaCl to water to create an aqueous salt solution, the following occurs:

I. Boiling point increases

II. Freezing point decreases

III. Vapor pressure decreases

A. I only

B. I and II only

C. II only

D. I, II, and III

23. When added to water, which of the following salts will create a basic solution?

A. NaCl

B. NaC$_2$H$_3$O$_2$

C. KI

D. NaNO$_3$

24. Three experiments were performed and the following results were obtained regarding the product formation rates.

$$A + B \rightarrow C$$

Experiment	[A]	[B]	Reaction Rate (mol/L–sec)
1	0.2	0.3	2.0 × 10^{-2}
2	0.4	0.3	8.0 × 10^{-2}
3	0.2	0.6	4.0 × 10^{-2}

What is the rate law for the reaction?

A. Rate = $k[A]^2$

B. Rate = $k[A]^2[B]$

C. Rate = $k[A][B]^2$

D. Rate = $k[B]^2$

25. If the following reaction produces 153 g of ammonia, how much heat is generated?

$$N_2(g) + 3H_2(g) \rightarrow 2NH_3(g) \quad \Delta H = -91.8 \text{ kJ}$$

A. 45.9 kJ

B. 183.6 kJ

C. 413 kJ

D. 826.2 kJ

180 ♦ CRACKING THE GOLDEN STATE EXAMINATION: CHEMISTRY

26. According to the following reaction, sulfuric acid reacted with sodium hydroxide will produce water and sodium sulfate.

$$H_2SO_4(aq) + 2NaOH(aq) \rightarrow 2H_2O(l) + Na_2SO_4(aq)$$

How many milliliters of 0.5 M NaOH will be needed to react completely with 50 mL of a 0.25 M H_2SO_4 solution?

A. 0.05 mL
B. 12.5 mL
C. 25 mL
D. 50 mL

27. According to Hund's rule, which of the following is true?

A. In order to obtain the lowest energy arrangement, electrons fill separate orbitals of the same subshell before pairing up with electrons.
B. In order to obtain the greatest energy arrangement, electrons fill separate orbitals of the same subshell before pairing up with electrons.
C. In order to obtain the lowest energy arrangement, electrons are paired within orbitals of the same subshell before filling other vacant orbitals.
D. In order to obtain the greatest energy arrangement, electrons are paired within orbitals of the same subshell before filling other vacant orbitals.

28. Which of the following statements CANNOT be true regarding two different carbon atoms:

A. their atomic weight can be different.
B. the number of protons in their nuclei can be different.
C. the number of neutrons in their nuclei can be different.
D. their mass numbers can be different.

29. An ideal gas that is 300 L in volume is prepared at a pressure of 380 mm Hg and a temperature of 225 °C. If the gas is placed in a container at high pressure, what will the volume of the gas when the container cools to a temperature of 25 °C and a pressure of 40 atm?

A. 0.44 L
B. 2.25 L
C. 6.25 L
D. 25.0 L

30. Which of the following elements has the greatest electronegativity?

A. Cs
B. K
C. F
D. P

WRITTEN RESPONSE QUESTION

Mercury barometers are used to measure atmospheric pressure. A barometer contains a glass tube that sits in a pool of mercury. Mercury contains properties that allow the substance to expand and rise at various pressures. If a capillary tube were placed in a pool of mercury, the level of mercury in the capillary would actually be lower than the level outside the tube.

Generally describe the process of measuring atmospheric pressure with a barometer and how the level of mercury is effected by pressure.

Explain why the level of mercury in a capillary is lower than the mercury outside the glass tube.

The Princeton Review

YOUR NAME: _____
(Print)   Last   First   M.I.

SIGNATURE: _____ DATE: ___/___/___

HOME ADDRESS: _____
(Print)   Number and Street

_____
         City   State   Zip Code

PHONE NO.: _____
(Print)

Completely darken bubbles with a No. 2 pencil. If you make a mistake, be sure to erase mark completely. Erase all stray marks.

Practice Test Two

1. Ⓐ Ⓑ Ⓒ Ⓓ
2. Ⓕ Ⓖ Ⓗ Ⓙ
3. Ⓐ Ⓑ Ⓒ Ⓓ
4. Ⓕ Ⓖ Ⓗ Ⓙ
5. Ⓐ Ⓑ Ⓒ Ⓓ
6. Ⓕ Ⓖ Ⓗ Ⓙ
7. Ⓐ Ⓑ Ⓒ Ⓓ
8. Ⓕ Ⓖ Ⓗ Ⓙ
9. Ⓐ Ⓑ Ⓒ Ⓓ
10. Ⓕ Ⓖ Ⓗ Ⓙ
11. Ⓐ Ⓑ Ⓒ Ⓓ
12. Ⓕ Ⓖ Ⓗ Ⓙ
13. Ⓐ Ⓑ Ⓒ Ⓓ
14. Ⓕ Ⓖ Ⓗ Ⓙ
15. Ⓐ Ⓑ Ⓒ Ⓓ
16. Ⓕ Ⓖ Ⓗ Ⓙ
17. Ⓐ Ⓑ Ⓒ Ⓓ
18. Ⓕ Ⓖ Ⓗ Ⓙ
19. Ⓐ Ⓑ Ⓒ Ⓓ
20. Ⓕ Ⓖ Ⓗ Ⓙ
21. Ⓐ Ⓑ Ⓒ Ⓓ
22. Ⓕ Ⓖ Ⓗ Ⓙ
23. Ⓐ Ⓑ Ⓒ Ⓓ
24. Ⓕ Ⓖ Ⓗ Ⓙ
25. Ⓐ Ⓑ Ⓒ Ⓓ
26. Ⓕ Ⓖ Ⓗ Ⓙ
27. Ⓐ Ⓑ Ⓒ Ⓓ
28. Ⓕ Ⓖ Ⓗ Ⓙ
29. Ⓐ Ⓑ Ⓒ Ⓓ
30. Ⓕ Ⓖ Ⓗ Ⓙ

Multiple-Choice Answer Key

1. D
2. B
3. B
4. A
5. C
6. C
7. B
8. A
9. D
10. C
11. D
12. B
13. B
14. A
15. B
16. B
17. D
18. C
19. D
20. D
21. A
22. D
23. B
24. B
25. C
26. D
27. A
28. B
29. B

Practice Test Two Answers and Explanations

1. **D.**

   Calcium's electrons are all spin-paired to complete the 1s, 2s, 2p, 3s, 3p, and 4s subshells. The other choices represent paramagnetic elements whose electrons are not all spin paired.

2. **B.**

   Rubidium has one additional electron than krypton, this electron is located in the 5s subshell.

3. **B.**

   Since N_2H_4 is on the right side of the reaction, multiply the first equation by −1.
   $N_2(g) + 2H_2O(l) \rightarrow N_2H_2(l) + O_2(g) =$
   $-1(-622) = 622$ kJ

   Since there are two H_2 on the left side of the reaction, multiply the second equation by 2.
   $2H_2(g) + O_2(g) \rightarrow 2H_2O(l) = 2(-286)$
   $= -572$ kJ

   Add the two equations:

   $N_2(g) + 2H_2O(l) \rightarrow N_2H_4(l) + O_2(g)$
   $\Delta H = 622$ kJ

   $2H_2(g) + O_2(g) \rightarrow 2H_2O(l) \quad \Delta H = -572$ kJ

   $N_2(g) + 2H_2(g) \rightarrow N_2H_4(l) \quad \Delta H = 50$ kJ

   Since ΔH is positive, energy has been absorbed. This means that the reaction is endothermic.

4. **A.**

   An sp hybrid orbital will form a linear molecule.

5. **C.**

   Increasing volume will disrupt the equilibrium causing the reaction to shift towards the side that produces more moles. In this reaction, the left side of the reaction produces more moles of gas. Answer choices A, B and D would cause the reaction to shift to the right.

6. **C.**

   Convert grams into moles:

   $2.8 \text{g N} \left(\dfrac{1 \text{ mol}}{14 \text{ g}} \right) = 0.2$ mol N

   $6.4 \text{g O} \left(\dfrac{1 \text{ mol}}{16 \text{ g}} \right) = 0.4$ mol O

   The ratio of N to O is 1 to 2; therefore, the empirical formula is NO_2.

7. **B.**

Try each answer choice until you find the one that works. If pressure is decreased and temperature is increased, choice B is possible as shown in the diagram below.

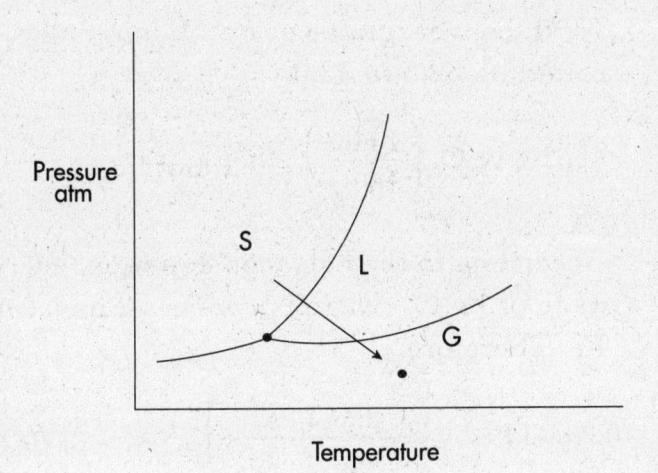

8. **A.**

The total number of valence electrons in Cl_2 is 14. $[(2)(7e^-)(Cl) = 14e^-]$ You can cross off choice B since it has only 12 electrons shown and answer choice D since it has 16 electrons shown. Choice C has 14 total electrons depicted, however, there are 10 electrons around each chlorine atom. According to the octet rule, atoms generally have a maximum of 8 electrons in their valence shell (2 electron for hydrogen atoms). Choice A depicts a total of 14 electrons with 8 electrons around each chlorine atom.

9. **D.**

First, convert the terms into the appropriate units:

Pressure: $2280 \text{ mmHg} \left(\dfrac{1 \text{ atm}}{760 \text{ mmHg}} \right) = 3$ atm

Volume: $90 \text{ mL} \left(\dfrac{1 \text{ L}}{1000 \text{ mL}} \right) = 0.090 \text{ L}$

Temperature: $50\,°C + 273 = 323 \text{ K}$

$PV = nRT$

$n = \dfrac{PV}{RT} = \dfrac{(3)(0.090)}{(0.08)(323)} = .0104 \text{ mol}$

Don't forget to convert moles of N_2 gas into grams:

$.0104 \text{ moles} \left(\dfrac{28 \text{ g}}{1 \text{ mol}} \right) \approx .3 \text{ grams}$

Watch out for choice B, which represents the number of moles and not grams.

10. **C.**

Statement I is not correct. Since there are two lone pairs of electrons present around the oxygen atom, the molecular shape of a water molecule is bent and not linear. Statement II does apply. Since water is a polar molecule, it will experience hydrogen bonding between the slightly positive hydrogen atom of one water molecule and the slightly negative oxygen atom of another water molecule. Statement III also applies since the orbitals on oxygen are sp^3 hybridized. The overlap of hydrogen's $1s$ orbital with a sp^3 hybrid oxygen orbital forms the O–H bond. A process of elimination technique can be used with this question. Once you have determined statement I to be incorrect, you can cross off choices A and D. Since statement II appears in both of the remaining choices, it has to be a true statement. Therefore, do not even look at statement II and just proceed to determine if statement III is correct.

11. **D.**

$$\text{Moles} = \frac{\text{grams}}{\text{MW}}$$

$$\text{Moles of NaNO}_3 = \frac{170 \text{ g}}{85 \text{ g/mol}} = 2 \text{ moles}$$

Every mole of $NaNO_3$ contains 3 moles of oxygen.

$$6 \text{ mol O} \left(\frac{16 \text{ g}}{\text{mol}} \right) = 96 \text{ g}$$

12. **B.**

First calculate the molecular weight of Fe_2O_3.

Molecular weight of Fe_2O_3 = 2(56) + 3(16) = 160 g/mol

Next, convert grams of Fe_2O_3 into moles, rounding 32.3 to 32.0:

$$32 \text{ g Fe}_2\text{O}_3 \left(\frac{1 \text{ mol}}{160 \text{ g}} \right) = 0.2 \text{ mol Fe}_2\text{O}_3$$

According to the balanced equation, every mole of Fe_2O_3 reacted produces 2 moles of Fe. Therefore:

$$0.2 \text{ mol Fe}_2\text{O}_3 \left(\frac{2 \text{ mol Fe}}{1 \text{ mol Fe}_2\text{O}_3} \right) = 0.4 \text{ mol Fe}$$

Finally, convert moles of Fe into grams:

$$0.4 \text{ mol Fe} \left(\frac{56 \text{ g}}{1 \text{ mol}} \right) = 22.4 \text{ g}$$

Knowing that we have used approximation, the answer closest to 22.4 g is choice B.

13. B.

First calculate the molecular weight of $HC_2H_3O_2$:

Molecular weight = 4(1) + 2(12) + 2(16) = 60 g/mol

Next, covert grams of acetic acid into moles:

$540 \text{ g } HC_2H_3O_2 \left(\dfrac{1 \text{ mol}}{60 \text{ g}} \right) = 9 \text{ mol } HC_2H_3O_2$

Be careful—every mole of $HC_2H_3O_2$ contains 4 moles of hydrogen.

Therefore, $36 \text{ mol H} \left(\dfrac{1 \text{ g}}{1 \text{ mol}} \right) = 36 \text{ g H}$

14. A.

Pressure and volume are inversely proportional. If you increase volume, pressure decreases. If you use process of elimination to cross off the choices where pressure is not decreased, you will be left with only one answer: choice A.

$$P_1V_1 = P_2V_1$$

$$P_2 = \dfrac{P_1V_1}{V_2} = \dfrac{(2)(2)}{8} = 0.5 \text{ atm}$$

15. B.

If the original nuclide emits 1 alpha particle, it emits something similar to an He ion ($^{4}_{2}He$). The mass number will be reduced by 4 and the atomic number will be reduced by 1. The beta particle, which is equivalent to an electron, will result in no change in the mass number and an increase in the atomic number by one for each beta particle.

$$^{210}_{82}Pb - ^{4}_{2}\alpha - ^{0}_{-1}\beta = ^{206}_{82}Pb$$

16. B.

Graham's law tells us that the gas with the greatest average speed will leak out of a small hole the quickest. Since kinetic energy is the same for all the gases ($KE = \dfrac{1}{2}mv^2$), as mass increases, velocity decreases. Therefore, the heaviest molecule, O_2, will have the lowest rate of effusion.

17. D.

To bring a 1-g sample of water from −10 °C to 110 °C, two phase changes occur. Therefore, we need to take into account the calories required to complete the phase change. The heat of fusion is equal to 80 cal/g, the amount of heat required for the phase change from ice to liquid water. The heat of vaporization is equal to 540 cal/g, the amount of heat required for the phase change from liquid water to steam. We already have a total of 620 cal of heat required just for the phase changes. The answer must be greater than 620. Looking at the choices, only D is greater than 620.

18. C.

The boiling point of a substance is reached when the phase change between a liquid and a gas occurs.

19. D.

Boiling point elevation

$\Delta T = k_b m i = (0.5)(3)(3) = 4.5\ °C$

The boiling point of water is 100 °C.

The boiling point of the $CaCl_2$ aqueous solution is then,

100 + 4.5 = 104.5 °C

20. D.

$$\text{Molality} = \frac{\text{moles of solute}}{\text{kilogram of solvent}}$$

First calculate the molecular weight of glucose,

Molecular weight $C_6H_{12}O_6$ = 6(12) + 12(1) + 6(16) = 180 g/mol

Next, convert grams of glucose into moles:

$$9\ g \left(\frac{1\ mol}{180\ g}\right) = 0.05\ mol\ C_6H_{12}O_6$$

Then put moles over kg of solutions:

$$\frac{.05}{.02} = 2.5\ m$$

Be careful: remember, to convert grams of water into kilograms:

$$20\ g \left(\frac{1\ kg}{1000\ g}\right) = 0.02\ kg$$

21. A.

If we increase the concentration of H_2 then equilibrium is disrupted and the reaction will move in the forward direction to use up the excess H_2.

22. D.

By adding NaCl to water, Na^+ and Cl^- ions dissociate to create an aqueous salt solution. These particles will interfere with the phase changes. The boiling point increases (boiling point elevation: $\Delta T = k_b m i$), freezing point decreases (freezing point depression: $\Delta T = k_f m i$), and vapor pressure decreases ($P = xP°$).

23. B.

A solution would be basic if the salt added was composed of the conjugate from a strong base (Na^+ from NaOH) and the conjugate from a weak acid ($C_2H_3O_2^-$ from $HC_2H_3O_2$). The other answer choices are salts composed of conjugates from strong acids and bases and would produce neutral solutions.

24. B.

Comparing experiments 1 and 2, when the concentration of [B] is held constant and the concentration of [A] is doubled, the rate of reaction is quadrupled. The reaction is therefore second order with respect to [A]: $[A]^2$. Comparing experiments 1 and 3, when the concentration of [A] is held constant and the concentration of [B] is doubled, the reaction also doubles. The reaction is first order with respect to [B]: [B]. The rate law is then = $k[A]^2[B]$.

25. **C.**

First convert grams of NH₃ into moles:

$$153 \text{ g NH}_3 \left(\frac{1 \text{ mol}}{17 \text{ g}}\right) = 9 \text{ mol NH}_3$$

According to the reaction, two moles of NH₃ are produced per 91.8 kJ of heat released.

Therefore:

$$9 \text{ mole NH}_3 \left(\frac{90 \text{ kJ}}{2 \text{ mol}}\right) \approx 405 \text{ kJ}$$

26. **D.**

First convert the 50 mL of 0.25 M H₂SO₄ into moles:

$$M = \frac{\text{mol}}{\text{L}}$$

mol = M × L = 0.25 M × 0.050 L = 0.0125 mol H₂SO₄

According to the balanced equation, every mole of H₂SO₄ reacts with 2 moles of NaOH. Therefore:

$$0.0125 \text{ mol H}_2\text{SO}_4 \left(\frac{2 \text{ mol NaOH}}{1 \text{ mol H}_2\text{SO}_4}\right) = 0.025 \text{ mol}$$

Finally, calculate how many liters of a 0.5 M NaOH solution equates to 0.025 mol of NaOH.

$$M = \frac{\text{mol}}{\text{L}}$$

$$L = \frac{\text{mol}}{M} = \frac{0.025 \text{ mol}}{0.5 \, M} = 0.050 \text{ L} = 50 \text{ mL}$$

Watch out! Choice A is the answer in a form that isn't converted to milliliters.

27. **A.**

This is the definition of Hund's rule. Electrons will fill separate orbitals of the same subshell, as follows:

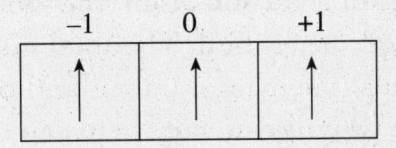

−1	0	+1
↑	↑	↑

Before spin pairing as follows:

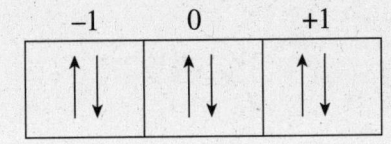

−1	0	+1
↑↓	↑↓	↑↓

28. **B.**

The total number of protons for all carbon atoms will be equal to 6. If somehow a proton were added or removed from the nucleus of a carbon atom, the atom would no longer be carbon. The total number of neutrons, the mass number, and the atomic weight can vary between carbon atoms.

29. **B.**

$$\frac{P_1 V_1}{T_1} = \frac{P_2 V_2}{T_2}$$

First convert all terms into the appropriate units:

$P_1 = \dfrac{(380 \text{ mmHg})(1 \text{ atm})}{760 \text{ mmHg}} = 0.5 \text{ atm}$

$V_1 = 300 \text{ L}$

$T_1 = 225\,°\text{C} + 273 = 498 \text{ K} \approx 500 \text{K}$

$P_2 = 40 \text{ atm}$

$T_2 = 25\,°\text{C} + 273 = 298 \text{K} \approx 300 \text{K}$

$V_2 = \dfrac{P_1 V_1 T_2}{P_2 T_1} = \dfrac{(0.5)(300)(300)}{(40)(500)} = 2.25 \text{ L}$

WRITTEN REPONSE QUESTION

Generally speaking, electronegativity increases as you move across a period (horizontal row) and decreases as you move down a group (vertical column).

Air pressure is dependent on altitude. This is a result of air exerting a downward pressure that will be sent through the pool of mercury and causing an upward pressure. At sea level, the mercury will rise to 760 mm above the pool of mercury. The height of the mercury and pressure are directly proportional, which means as pressure increases, so does the level of mercury.

Liquids rising in a capillary is related to surface tension. Mercury will not rise in a capillary and will have a convex meniscus because mercury is not attracted to the glass.

Chapter 18

PRACTICE TEST THREE

1. Which of the following is true regarding an element whose electron configuration is $1s^2 2s^2 2p^6 3s^2 3p^6 4s^2 3d^{10} 4p^5$?

   A. Noble gas
   B. Most stable in its ground state
   C. A transition metal
   D. A halogen

2. If we add sodium chloride to a saturated aqueous solution of nickel chloride, a precipitate will appear. What is the precipitate?

   A. Nickel chloride
   B. Sodium chloride
   C. Chlorine
   D. Nickel

3. Which of the following describes the geometrical shape of NH_3?

   A. Bent
   B. Linear
   C. Trigonal pyramidal
   D. Tetrahedral

4. A scientist creates 10 L of 2 M NaOH. If she wants to later dilute the solution to 0.1 M, how much water must she add?

   A. 2 L
   B. 10 L
   C. 190 L
   D. 200 L

5. If 200 g of calcium oxide react with 200 g of carbon as depicted in the reaction below, which is the limiting reagent?

   $$CaO(s) + 3C(s) \rightarrow CaC_2(s) + CO(g)$$

   A. $CaO(s)$
   B. $C(s)$
   C. $CaC_2(s)$
   D. $CO(g)$

6. How many calories of heat must be removed to bring a 1 g sample of water from 130 °C to 80 °C? (Heat of fusion for water = 80 cal/g, heat of vaporization for water = 540 cal/g, specific heat of ice and steam = 0.5 cal/g °C, specific heat of liquid water = 1 cal/g °C.)

   A. 50
   B. 115
   C. 540
   D. 575

7. If a scientist wanted to lower the pH of 1 L of an aqueous NaOH solution from pH = 13 to pH = 11, how much distilled water would she need to add?

   A. 9 L
   B. 10 L
   C. 99 L
   D. 100 L

8. If $^{8}_{5}B$ undergoes positron emission, which of the following nuclides will result?

A. $^{8}_{4}Be$

B. $^{7}_{4}Be$

C. $^{8}_{6}C$

D. $^{9}_{6}C$

9.

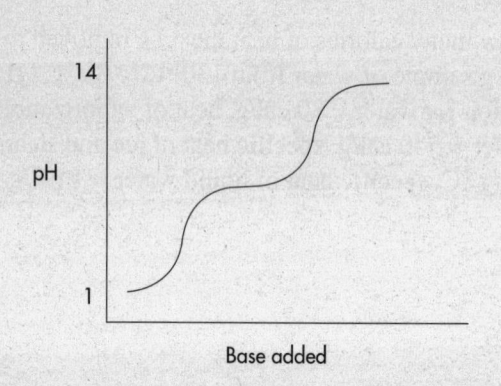

[Graph: pH vs Base added, y-axis from 1 to 14, showing titration curve]

The diagram above could represent which of the following?

A. The addition of NaOH to HCl
B. The addition of HCl to NaOH
C. The addition of $Ba(OH)_2$ to H_2SO_4
D. The addition of H_2SO_4 to $Ba(OH)_2$

10. What is the empirical formula of an 8.4 g sample of a compound that is comprised of 3.6 g of carbon and the rest oxygen?

A. CO
B. CO_2
C. C_2O
D. CO_4

11. $A + 2B \rightarrow 2C + D$

Experiment	[A]	[B]	Reaction Rate (mol/L × sec)
1	0.2	0.1	4.0×10^{-3}
2	0.2	0.2	8.0×10^{-3}
3	0.4	0.1	1.2×10^{-2}

Three experiments were performed and the following results were obtained regarding the product formation rates. What is the rate law for the reaction?

A. Rate = $k[A]^2$
B. Rate = $k[A]^2[B]$
C. Rate = $k[A]^2[B]^2$
D. Rate = $k[A]^2[B]^2[C]$

12. $A + B + C \rightarrow D + E + C$

In the reaction listed above, C can be considered which of the following?

I. A reactant
II. A product
III. A catalyst

A. I only
B. I and II only
C. III only
D. I, II, and III

198 ◆ CRACKING THE GOLDEN STATE EXAMINATION: CHEMISTRY

13. In the Haber process, ammonia (NH$_3$) is produced from the reaction of atmospheric nitrogen (N$_2$) and hydrogen (H) at high temperature and pressure.

$$N_2(g) + 3H_2(g) \rightarrow 2NH_3$$

If we start with 7.0 g of atmospheric nitrogen, how many grams of ammonia will be produced?

A. 2.1 g
B. 4.2 g
C. 8.5 g
D. 17.0 g

14. If 250 g of calcium carbonate (CaCO$_3$) reacted according to the following equation at 0 °C and 1 atm, what volume of carbon dioxide (CO$_2$) gas would be produced?

$$CaCO_3 \rightarrow CaO + CO_2$$

($R = 0.0821$ L atm/mol × K)

A. 22 L
B. 34 L
C. 45 L
D. 56 L

15. An ideal gas occupies an 800 mL container at 25 °C and has a pressure of 380 mmHg. How many moles of gas are present? ($R = 0.0821$ L × atm/mol × K)

A. $\frac{1}{60}$ mol

B. $\frac{1}{6}$ mol

C. 1 mol
D. 6 mol

16. Which of the following molecules is not formed by covalent bonds?

A. NaCl
B. Cl$_2$
C. HCl
D. CH$_4$

17. A 5 L sample of an oxygen and nitrogen gas mixture contains 16 g of O$_2$ and 28 g of N$_2$. At 25 °C, what is the total pressure of the gases?
($R = 0.0821$ L · atm/mol · K)

A. 2.4 atm
B. 4.8 atm
C. 7.2 atm
D. 14.4 atm

18. A gas mixture contains O$_2$, CO$_2$, N$_2$, and H$_2$. Which of the following represents the order of decreasing average molecular speed?

A. CO$_2$, N$_2$, O$_2$, H$_2$
B. H$_2$, N$_2$, O$_2$, CO$_2$
C. CO$_2$, O$_2$, N$_2$, H$_2$
D. H$_2$, O$_2$, N$_2$, CO$_2$

19. A balloon contains a mixture of gases at STP. If 0.5 mol of oxygen gas and 0.5 mol of hydrogen gas are present which of the following is true?

   I. Both gases have the same average kinetic energy.

   II. Both gases have the same partial pressures.

   III. Both gases have the same average molecular speeds.

   A. I only
   B. I and II only
   C. II and III only
   D. I, II, and III

20.

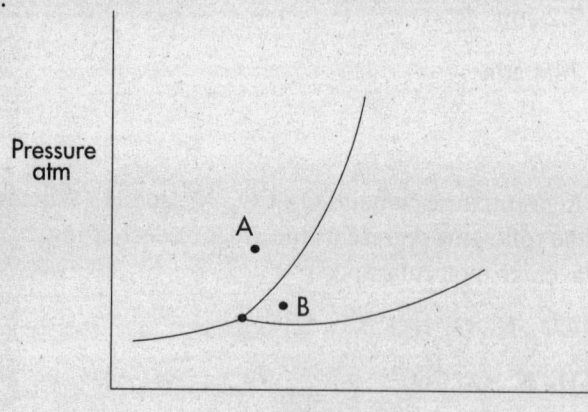

   What phase change occurs from point A to B when pressure is decreased and temperature is held constant?
   A. melting
   B. vaporization
   C. sublimation
   D. freezing

21. When pressure is decreased and temperature is held constant, which of the following phase changes can occur?

   I. Melting
   II. Vaporization
   III. Freezing

   A. I only
   B. I and II only
   C. II only
   D. I, II, and III

22. Which of the following expressions represents the change in temperature of a 500 g sample of silver when 50 cal are added to the sample? (Specific heat of silver = 0.06 cal/g °C)

   A. $\dfrac{(50)}{(500)(0.06)}$

   B. $(50)(500)(0.06)$

   C. $\dfrac{(500)(0.06)}{50}$

   D. $\dfrac{(50)(500)}{(0.06)}$

23. In order to create a 2 M solution of HCl from 4 L of a 5 M solution, how many liters of water must be added?

   A. 4 L
   B. 6 L
   C. 10 L
   D. 14 L

24. Of the following listed aqueous solutions, which one will have the lowest freezing point?

A. 1 m NaCl

B. 1 m KCl

C. 1 m $C_6H_{12}O_6$

D. 1 m $CaCl_2$

25. A gas mixture created from the following equation is at equilibrium.

$$CO(g) + 3H_2(g) \leftrightarrow CH_4(g) + H_2O(g)$$

If the equilibrium constant is equal to 3.9 and 0.5 mol of CO, 0.20 mol H_2, and 0.01 mol of CH_4 are present, how many moles of H_2O are present in the gaseous mixture?

A. $\dfrac{(0.01)}{(3.9)(0.5)(0.2)^3}$

B. $\dfrac{(3.9)(0.5)(0.2)^3}{(0.01)}$

C. $\dfrac{(0.5)(0.2)^3}{(3.9)(0.01)}$

D. $\dfrac{(0.01)(3.9)}{(0.5)(0.2)^3}$

26. A solution is comprised of equal masses of 2 substances, where substance A has a molecular weight of 200 g/mol and substance B has a molecular weight of 50 g/mol. What is the mole fraction of substance A?

A. $\dfrac{1}{5}$

B. $\dfrac{1}{4}$

C. $\dfrac{3}{4}$

D. $\dfrac{4}{5}$

27. $2NO_2(g) \leftrightarrow N_2O_4(g)$

If pressure is increased to the above equilibrium expression keeping temperature constant, which of the following will occur?

A. The reaction will shift to the left.

B. The reaction will shift to the right.

C. There will be no effect.

D. The volume will also increase.

28. Which of the following is a strong electrolyte?

A. $C_6H_{12}O_6$

B. HCl

C. H_2O

D. CH_4

29. Which of the following statements accurately describes the reaction depicted in the diagram below?

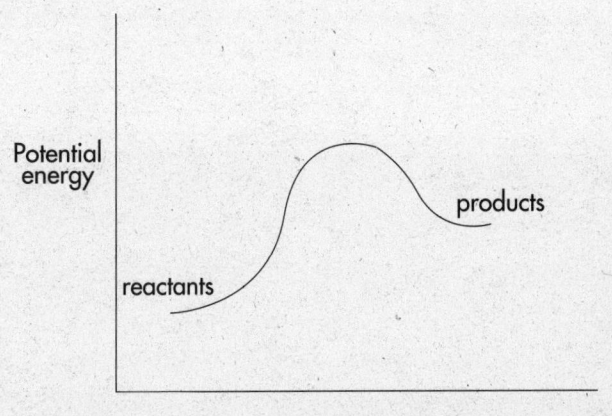

A. This is an endothermic reaction where energy is absorbed.

B. This is an exothermic reaction where energy is absorbed.

C. This is an endothermic reaction where energy is released.

D. This is an exothermic reaction where energy is released.

30. Hydrogen gas is reacted with oxygen to yield water vapor.

$$2H_2(g) + O_2(g) \rightarrow 2H_2O(g) \quad \Delta H = -484 \text{ kJ}$$

What is the ΔH for each gram of hydrogen?

A. −121 kJ

B. −242 kJ

C. 121 kJ

D. 242 kJ

31. Which set of four quantum numbers (n, l, m, s) most likely represents the highest energy valence electron in a carbon atom in its ground state?

A. 2, 0, −1, +1/2

B. 2, 0, 0, +1/2

C. 2, 1, −1, +1/2

D. 2, 1, 0, +1/2

WRITTEN RESPONSE QUESTION

Hot air ballooning can be used as a mode of transportation or as a leisure activity. Propane gas is the main fueling agent used in hot air ballooning. Explain why a hot air balloon rises.

The Princeton Review

YOUR NAME: _____
(Print)   Last   First   M.I.

SIGNATURE: _____ DATE: ___/___/___

HOME ADDRESS: _____
(Print)   Number and Street

_____
City   State   Zip Code

PHONE NO.: _____
(Print)

Completely darken bubbles with a No. 2 pencil. If you make a mistake, be sure to erase mark completely. Erase all stray marks.

Practice Test Three

1. Ⓐ Ⓑ Ⓒ Ⓓ
2. Ⓕ Ⓖ Ⓗ Ⓙ
3. Ⓐ Ⓑ Ⓒ Ⓓ
4. Ⓕ Ⓖ Ⓗ Ⓙ
5. Ⓐ Ⓑ Ⓒ Ⓓ
6. Ⓕ Ⓖ Ⓗ Ⓙ
7. Ⓐ Ⓑ Ⓒ Ⓓ
8. Ⓕ Ⓖ Ⓗ Ⓙ
9. Ⓐ Ⓑ Ⓒ Ⓓ
10. Ⓕ Ⓖ Ⓗ Ⓙ
11. Ⓐ Ⓑ Ⓒ Ⓓ
12. Ⓕ Ⓖ Ⓗ Ⓙ
13. Ⓐ Ⓑ Ⓒ Ⓓ
14. Ⓕ Ⓖ Ⓗ Ⓙ
15. Ⓐ Ⓑ Ⓒ Ⓓ
16. Ⓕ Ⓖ Ⓗ Ⓙ
17. Ⓐ Ⓑ Ⓒ Ⓓ
18. Ⓕ Ⓖ Ⓗ Ⓙ
19. Ⓐ Ⓑ Ⓒ Ⓓ
20. Ⓕ Ⓖ Ⓗ Ⓙ
21. Ⓐ Ⓑ Ⓒ Ⓓ
22. Ⓕ Ⓖ Ⓗ Ⓙ
23. Ⓐ Ⓑ Ⓒ Ⓓ
24. Ⓕ Ⓖ Ⓗ Ⓙ
25. Ⓐ Ⓑ Ⓒ Ⓓ
26. Ⓕ Ⓖ Ⓗ Ⓙ
27. Ⓐ Ⓑ Ⓒ Ⓓ
28. Ⓕ Ⓖ Ⓗ Ⓙ
29. Ⓐ Ⓑ Ⓒ Ⓓ
30. Ⓕ Ⓖ Ⓗ Ⓙ
31. Ⓐ Ⓑ Ⓒ Ⓓ

Multiple-Choice Answer Key

1. D
2. A
3. C
4. C
5. A
6. D
7. C
8. A
9. C
10. A
11. B
12. C
13. C
14. D
15. A
16. A
17. C
18. B
19. B
20. A
21. D
22. A
23. B
24. D
25. B
26. A
27. B
28. B
29. A
30. A
31. D

Practice Test Three Answers and Explanations

1. **D.**

   The total number of electrons is 35. If we look at the periodic table, we can see that 35 is bromine, which is in group VII (the halogen group). The elements in this group are not stable in their ground state because their valence shell is not completely full.

2. **A.**

   By introducing additional chlorine ions into solution, the solution's equilibrium is disrupted and shifts the direction of the reaction towards the reactants. Since sodium chloride is more stable than nickel chloride, the nickel chloride will precipitate out of solution.

3. **C.**

   Nitrogen has five valence electrons and each hydrogen has one, resulting in a total of eight valence electrons for a molecule of NH_3. Using a Lewis dot structure, we can arrange the eight valence electrons around the central nitrogen atom and the three hydrogen atoms.

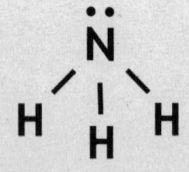

   $$\overset{..}{N}\begin{matrix} \\ H \quad H \\ H \end{matrix}$$

   The central nitrogen atom contains three bonds and a lone pair of electrons. The hydrogen atoms and the lone pair of electrons will distribute themselves as far apart as possible from each other to minimize the electron repulsions. The resulting shape is trigonal pyramidal.

4. **C.**

   First calculate the number of moles of NaOH:

   $$\text{Molarity } (M) = \frac{\text{moles}}{\text{liter}}$$

   moles = molarity × liters = (2)(10) = 20 mol

   Keeping the number of NaOH moles the same, calculate the new volume of NaOH at 0.1 M:

   $$\text{liters} = \frac{\text{moles}}{\text{Molarity}} = (20)/(0.1) = 200 \text{ L}$$

   The total volume is 200 L. If the scientist started with 10 L, that means she needs 190 L to dilute her NaOH solution to 0.1 M.

5. **A.**

   The limiting reagent is the reactant that will run out first. You can first cross off choices C and D since they are the products and not the reactants. Three moles of C (s) are used in the reaction for every one mole of CaO (s). With 200 g of CaO (s) we can calculate the number of moles:

   $$200 \text{ g CaO}(s) \left(\frac{1 \text{ mol}}{56.08 \text{ g}}\right) = 3.57 \text{ mol CaO }(s)$$

   If we start with 3.57 mol of CaO (s) then we need only three times that amount of solid carbon, which is 10.71 mol.

208 ♦ CRACKING THE GOLDEN STATE EXAMINATION: CHEMISTRY

We can determine how many moles of solid carbon are available:

$$200 \text{ g } C(s) \left(\frac{1 \text{ mol}}{12.01 \text{ g}}\right) = 16.65 \text{ mol } C(s)$$

We therefore have excess solid carbon. This means the limiting reagent in the example is CaO (s).

6. **D.**

   Calculate in steps:
   $$130 \text{ °C} \rightarrow 100 \text{ °C}$$
   $$q = mc\Delta T = (1)(0.5)(30) = 15 \text{ cal}$$

   Phase change: gas $\rightarrow$ liquid = 540 cal
   $$100 \text{ °C} \rightarrow 80 \text{ °C}$$
   $$q = mc\Delta T = (1)(1)(20) = 20 \text{ cal}$$

   Total calories to be removed = 15 + 540 + 20 = 575 cal

   Be careful: The heat of fusion of water (80 cal/g) was given, but you did not need to use that value for this question.

7. **C.**

   pH = 13, therefore pOH = 1
   pH = 11, therefore pOH = 3
   pOH = 1 = p[OH⁻] = –log[0.1]
   pOH = 3 = p[OH⁻] = –log[0.001]

   moles = molarity × volume
   $$M_1V_1 = M_2V_2$$
   $$V_2 = \frac{M_1V_1}{M_2} = \frac{(0.1)(1)}{(0.001)} = 100 \text{ L}$$

   Be careful: If the total volume is 100 L, then the scientist added 99 L of water to the original 1 L of NaOH solution.

8. **A.**

   With positron emission, the nucleus changes a proton into a neutron and a positron (+) and emits the positron. The mass number will remain the same and the atomic number will decrease by one.

   $$^{8}_{5}B + ^{0}_{1}\beta = ^{8}_{4}Be$$

9. **C.**

   The diagram represents the titration of a diprotic acid by a base. The titration curve has 2 bumps to represent the two hydrogen ions that H_2SO_4 has to give up.

10. **A.**

    First calculate grams of oxygen:
    8.4 total grams – 3.6 g carbon = 4.8 g oxygen

    Next convert grams to moles:
    $$3.6 \text{ g carbon} \left(\frac{1 \text{ mol}}{12 \text{ g}}\right) = 0.3 \text{ mol}$$
    $$4.8 \text{ g oxygen} \left(\frac{1 \text{ mol}}{16 \text{ grams}}\right) = 0.3 \text{ mol}$$

    The ratio of carbon to oxygen if 1:1; therefore, the empirical formula is CO.

11. B.

First of all, note that all of the choices contain $[A]^2$. Therefore, we already know that the reaction is going to be second order with respect to [A]. We only need to look at [B]. Comparing experiments 1 and 2, when the concentration of [A] is held constant and the concentration of [B] is doubled, the rate of reaction is also doubled. The reaction is therefore first order with respect to [B]. The rate law is then $k[A]^2[B]$.

12. C.

C is a catalyst. It may have changed during intermediate steps but the final product is returned in its original form. When a substance is added as a reactant to speed up a reaction process but does not alter the final products since it is returned as a product itself, this is known as a catalyst.

13. C.

First convert grams of N_2 into moles:

$$7.0 \text{ g} \left(\frac{1 \text{ mol}}{28 \text{ g}} \right) = 0.25 \text{ mol } N_2$$

According to the balanced equation, every mole of N_2 reacted produces 2 moles of NH_3. Therefore:

$$0.25 \text{ mol } N_2 \left(\frac{2 \text{ mol } NH_3}{1 \text{ mol } N_2} \right) = 0.5 \text{ mol } NH_3$$

Finally, convert moles of NH_3 into grams:

$$0.5 \text{ mol } NH_3 \left(\frac{17 \text{ g}}{1 \text{ mol}} \right) = 8.5 \text{ g } NH_3$$

14. D.

First calculate the molecular weight of $CaCO_3$:

MW = (40) + (12) + 3(16) = 100 g/mol

Next, convert grams of $CaCO_3$ into moles:

$$250 \text{ g CaCO}_3 \left(\frac{1 \text{ mol}}{100 \text{ g}} \right) = 2.5 \text{ moles}$$

According to the balanced equation, every mole of $CaCO_3$ reacted produces 1 mole of CO_2 gas. Therefore:

2.5 mol $CaCO_3$ = 2.5 mol CO_2

Using the ideal gas law ($PV = nRT$), we can calculate the volume of CO_2 gas produced:

$$V = \frac{nRT}{P} = \frac{(2.5)(0.0821)(273)}{(1)} = 56 \text{ L}$$

STP is represented by 1 atm and 0 °C. You may recall that volume of a gas at STP is 22.4 L. The volume of CO_2 gas in this example would then be (2.5)(22.4) = 56 L

15. A.

First convert all the terms into the appropriate units:

$$\text{Volume} = 800 \text{ mL} \left(\frac{1 \text{ L}}{1000 \text{ mL}} \right) = 0.800 \text{ L}$$

Temperature: 25 °C + 273 = 298K ≈ 300K

$$\text{Pressure: } 380 \text{ mmHg} \left(\frac{1 \text{ atm}}{760 \text{ mmHg}} \right) = 0.5 \text{ atm}$$

$$PV = nRT$$

$$n = \frac{PV}{RT} = \frac{(0.5)(0.800)}{(0.08)(300)} = \frac{1}{60} \text{ mol}$$

16. **A.**

There is a large difference in electronegativity between sodium (Na) and chloride (Cl). Therefore, the bond between them will be ionic. Chlorine, having the larger electronegativity value, will steal the lone electron in sodium's valence shell to complete its octet. Even though there is a difference in electronegativity in HCl, the difference is not great enough for an ionic bond. The resulting bond is a polar covalent bond. Choices B and D are not even polar molecules.

17. **C.**

First convert grams into moles:

$$16 \text{ g O}_2 \left(\frac{1 \text{ mol}}{32 \text{ g}}\right) = 0.5 \text{ mol O}_2$$

$$PV = nRT$$

$$PO_2 = \frac{nRT}{V} = \frac{(0.5)(0.08)(300)}{(5)} = 2.4 \text{ atm}$$

$$28 \text{ g N}_2 \left(\frac{1 \text{ mol}}{28 \text{ g}}\right) = 1 \text{ mol N}_2$$

$$PN_2 = \frac{nRT}{V} = \frac{(1)(0.08)(300)}{(5)} = 4.8 \text{ atm}$$

$$P_{total} = PO_2 + PN_2 = 2.4 + 4.8 = 7.2 \text{ atm}$$

18. **B.**

According to Graham's law, gases within the same environment will all have the same kinetic energy ($KE = \frac{1}{2}mv^2$). Therefore, if molecules have a heavier mass, their velocity will be lower than a molecule of lighter mass. In order to list the molecules from greatest speed to slowest speed, you must list them from lightest to heaviest. The molecular weights of these molecules are as follows: $H_2 = 2$, $N_2 = 28$, $O_2 = 32$, $CO_2 = 44$.

19. **B.**

According to the kinetic molecular theory, gas molecules within the same environment will have the same average kinetic energy; therefore item I is correct. According to Dalton's law, the sum of the partial pressures of all the gases is equal to the total pressure of the mixture. Since pressure is proportional to moles, the mole fraction is also equal to the partial pressure. Therefore, item II is correct. We just proved that the average kinetic energy for all the gases present in the mixture is the same 9$KE = \frac{1}{2}mv^2$). Therefore, heavier molecules will move slower than lighter particles. Since oxygen gas is heavier than hydrogen gas, the oxygen molecules will move at a slower rate. Therefore, item III is incorrect.

20. **A.**

Moving from point A to point B as a result of decreasing pressure represents the change from a solid to a liquid (melting).

21. **D.**

All three phase changes can occur as pressure is increased and temperature is held constant. This is shown in the two diagrams below.

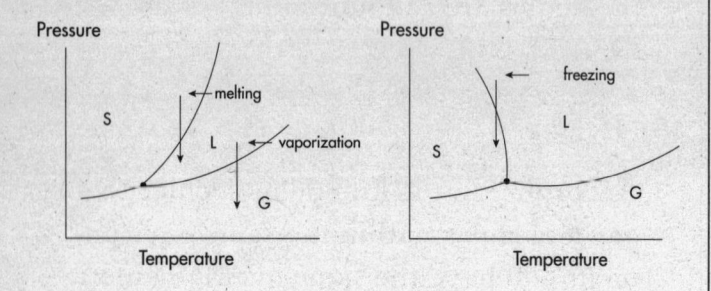

22. **A.**

$$q = mc\Delta T$$

$$\Delta T = \frac{q}{mc} = \frac{(50)}{(500)(0.06)}$$

23. **B.**

First calculate the number of moles of HCl.

$$\text{molarity} = \frac{\text{moles}}{\text{liter}}$$

moles = molarity × liters = (5)(4) = 20 mol

Keeping the number of HCl moles the same, calculate the new volume of HCl at 2 M.

$$\text{liters} = \frac{\text{moles}}{\text{molarity}} = \frac{20}{2} = 10 \text{ L}$$

The total volume is 10 L. If we started with 4 L, that means we must have added 6 L.

24. **D.**

Freezing point depression depends in part on the number of particles in solution. Choices A and B dissociate into 2 particles, choice C does not dissociate, and choice D dissociates into 3 particles. Since all of the solutions have the same molality, the greatest freezing point depression will be the solution that dissociates into the most ions (choice D).

25. **B.**

Equilibrium constant:

$$K = \frac{[CH_4][H_2O]}{[CO][H_2]^3}$$

$$[H_2O] = \frac{K[CO][H_2]^3}{[CH_4]}$$

26. **A.**

According to the question, the solution contains equal masses of substance A and substance B. Since the mole fraction is a fractional quantity, we can plug in any value for the mass. Let's assume 200 g of each substance. Convert grams to moles:

Substance A: $200g \left(\dfrac{1 \text{ mol}}{200 \text{ g}} \right) = 1$ mol

Substance B: $200g \left(\dfrac{1 \text{ mol}}{50g} \right) = 4$ mol

Total moles A + B = 1 + 4 = 5

Mole fraction of A = $\dfrac{1}{5}$

27. B.

According to the ideal gas equation ($PV = nRT$), pressure and volume are inversely proportional. If pressure is increased at a constant temperature, the volume will decrease, causing a strain on the reaction. The reaction will then shift to the side that produces fewer moles of gas to compensate for the smaller volume. In this reaction the right side of the equation has fewer moles of gas. Therefore, the reaction will shift to the right.

28. B.

An electrolyte is a substance that conducts electricity. A substance that dissociates into positive and negative ions will conduct electricity. The only molecule listed that will dissociate into ions is HCl.

29. A.

In the reaction depicted, the reactants are at a lower energy level than the products. Therefore, energy is being absorbed and the reaction is endothermic.

30. A.

In this reaction, 484 kJ of heat is released for every 2 moles of hydrogen reacted. Therefore:

$$\left(\frac{-484 \text{ kJ}}{2 \text{ mol H}_2}\right)\left(\frac{1 \text{ mol}}{2 \text{ g}}\right) = -121 \text{ kJ}$$

31. D

Carbon's electron configuration is $1s^2 2s^2 2p^2$. Its valence electrons are in the $2p$ subshell. Therefore $n = 2$, $l = p = 1$, $m = 0$, $s = +\frac{1}{2}$.

WRITTEN REPONSE QUESTION

Density is the measure of mass per volume.

$$\text{Density} = \frac{\text{mass}}{\text{volume}}$$

With gases, density is also dependent on temperature and pressure.

$$PV = nRT$$

$$V = \frac{nRT}{P}$$

The propane gas in a hot air balloon heats the air within the balloon. As temperature rises, so does volume. Therefore, the air expands to fill the hot air balloon.

Since volume and density are inversely proportional, density will decrease as volume increases, allowing the balloon to rise and float in the air.

Chapter 19

PRACTICE TEST FOUR

1. A syringe contains 50.0 mL of an ideal gas at 273 K. If the pressure was held constant and temperature was increased to 410 K, what is the new volume that the gas occupies?

   A. 12.5 mL
   B. 25 mL
   C. 50 mL
   D. 75 mL

2.

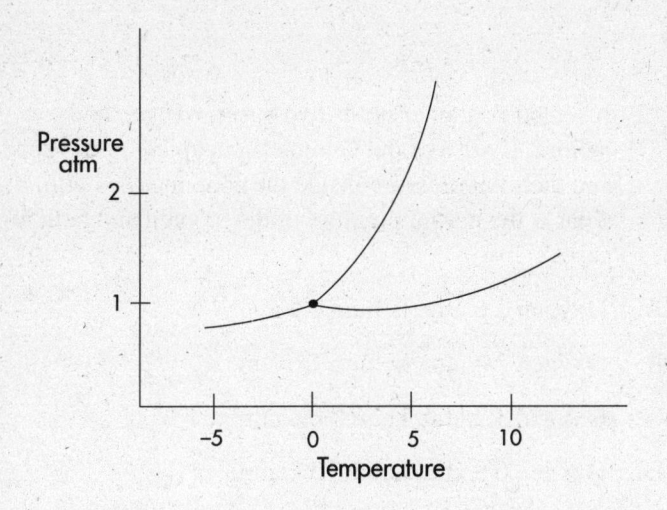

   According to the illustration above, at 10 °C and 2 atm, this substance would be in which following phase?

   A. Solid
   B. Liquid
   C. Gas
   D. Triple point

3. Which of the following order of elements represents increasing electronegativity?

   A. Li, C, N, F
   B. F, N, C, Li
   C. C, F, Li, N
   D. C, N, F, Li

4. According to the following reaction, if 300 mL of hydrogen gas is produced at 8 atm and 25 °C, how many grams of hydrogen gas are produced?

   $$2HCl(aq) + Zn(s) \rightarrow ZnCl_2(aq) + H_2(g)$$

   A. 0.1 g
   B. 0.2 g
   C. 1.0 g
   D. 2.0 g

5. What is the approximate change in temperature when 60 cal of heat are added to a 100 g sample of tungsten? (Specific heat of tungsten = 0.0321 cal/g °C)

   A. 0.2 °C
   B. 2 °C
   C. 20 °C
   D. 200 °C

6. Which of the following equilibrium equations would produce additional reactants and favor the reverse direction if volume was increased at a constant temperature?

   A. $2NO(g) + O_2(g) \leftrightarrow 2NO_2(g)$
   B. $CH_4(g) + 2H_2S(g) \leftrightarrow CS_2(g) + 4H_2(g)$
   C. $CO(g) + H_2O(g) \leftrightarrow CO_2(g) + H_2(g)$
   D. $H_2(g) + I_2(g) \leftrightarrow 2HI(g)$

7. Which of the following describes the geometrical shape of CO_2?

   A. Bent
   B. Linear
   C. Trigonal planar
   D. Tetrahedral

8. $2CH_3OH(g) + 3O_2(g) \rightarrow 2CO_2(g) + 4H_2O(g)$

   Based on the following data, what is the change in enthalpy for the above reaction?

Compound	ΔH
CH_3OH	−201 kJ
O_2	0
CO_2	−394 kJ
H_2O	−242 kJ

   A. −1354 kJ
   B. −837 kJ
   C. 837 kJ
   D. 1354 kJ

9. The strongest intermolecular attraction would be encountered by which of the following substances?

   A. BeF_2
   B. CO_2
   C. H_2O
   D. CH_4

10. If a compound contains 36% calcium and 64% chlorine, what is the empirical formula?

    A. CaCl
    B. Ca_2Cl
    C. $CaCl_2$
    D. Ca_2Cl_4

11. Hydrochloric acid reacted with silver nitrate will produce a white precipitate (silver chloride). The following illustrates this reaction:

    $HCl(aq) + AgNO_3(aq) \rightarrow HNO_3(aq) + AgCl(s)$

    If 50 mL of HCl reacted with $AgNO_3$ yields 35.75 g of the silver nitrate precipitate, what was the molarity of the HCl solution?

    A. 0.05 M
    B. 0.5 M
    C. 5 M
    D. 50 M

12. A sealed container holds two gases, oxygen and helium. If 40% of the volume is occupied by oxygen and the total pressure inside the container is 4 atm, what is the partial pressure of the oxygen and helium gases?

    A. Oxygen 1.6 atm, helium 2.4 atm
    B. Oxygen 2.4 atm, helium 1.6 atm
    C. Oxygen 3.2 atm, helium 0.8 atm
    D. Oxygen 0.8 atm, helium 3.2 atm

13. A sample of air contains 1.68 g of N_2 plus a mixture of other gases, including oxygen and carbon dioxide. This sample occupies 4 L at STP. What is the partial pressure of N_2 in the sample of air?

    A. 0.33 atm
    B. 0.66 atm
    C. 1 atm
    D. 1.33 atm

14. What is the percent composition by mass of sodium, oxygen, and hydrogen in NaOH?

    A. 33.3% Na, 33.3% O, 33.3% H
    B. 48% Na, 40% O, 12% H
    C. 57.5 % Na, 40% O, 2.5% H
    D. 57.5 % Na, 20% O, 22.5% H

15. Which of the following molecular shapes CANNOT be a nonpolar structure?

    A. linear
    B. bent
    C. tetrahedral
    D. square

16.

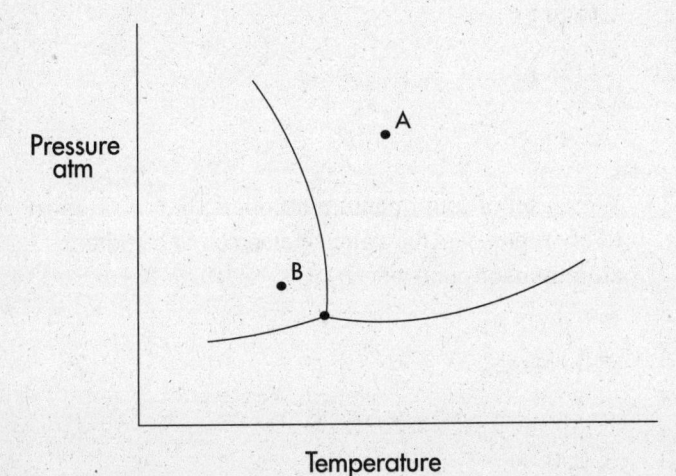

The above illustration represents the phase diagram for water. What phase change is occurring from point A to point B, when pressure is decreased and temperature is held constant?

    A. melting
    B. vaporization
    C. sublimation
    D. freezing

17. Which of the following expressions represents the mole fraction of glucose ($C_6H_{12}O_6$) in a 2.5 m aqueous solution?

    A. (2.5)/(1000)
    B. (2.5)/[(2.5) + (55.6)]
    C. (2.5) + (55.6)/(2.5)
    D. (2.5)/(55.6)

18. Which of the following expressions represents the equilibrium constant for the following reaction:

$$CO(g) + 2H_2(g) \leftrightarrow CH_3OH(g)$$

    A. $\dfrac{[CH_3OH]}{[CO][H_2]^2}$

    B. $\dfrac{[CO][H_2]^2}{[CH_3OH]}$

    C. $\dfrac{[CO][H_2]}{[CH_3OH]}$

    D. $\dfrac{[CH_3OH]}{[CO][H_2]}$

19. $KNO_3(s)$ + energy → $K^+(aq)$ + $NO_3^-(aq)$

    Which of the following would cause an increase in K^+ ions?

    I. An increase in temperature
    II. Adding additional KNO_3
    III. Adding additional NO_3^- ions

    A. I only
    B. I and II only
    C. II and III only
    D. I, II, and III

20. When added to water, which of the following salts would create a solution with pH less than 7?

A. $NaC_2H_3O_2$

B. NaCl

C. KNO_3

D. NH_4Cl

21. How many milliliters of 0.5 M HCl would be needed to neutralize 50 mL of 0.2 M NaOH?

A. 20 mL

B. 50 mL

C. 100 mL

D. 125 mL

22. A + B → C + D

Experiment	[A]	[B]	Reaction Rate (mol/L × sec)
1	0.1	0.2	6.0×10^{-3}
2	0.1	0.4	6.0×10^{-3}
3	0.2	0.2	1.2×10^{-2}

Three experiments were performed and the following results were obtained regarding the product formation rates. What is the rate law for the reaction?

A. Rate = k[A]

B. Rate = k[A]²

C. Rate = k[B]

D. Rate = k[A][B]

23. Considering each of the following aqueous solution to be 1 m, which one will boil at the highest temperature?

A. $Mg(OH)_2$

B. $C_6H_{12}O_6$

C. KCl

D. NaCl

24. $4Al(s) + 3MnO_2(s) \rightarrow 2Al_2O_3(s) + 3Mn(s)$

Using the following information what is the ΔH for the reaction of manganese dioxide listed above?

$$2Al(s) + \frac{3}{2}O_2(g) \rightarrow Al_2O_3(s) \quad \Delta H = -1676 \text{ kJ}$$

$$Mn(s) + O_2(g) \rightarrow MnO_2(s) \quad \Delta H = -521 \text{ kJ}$$

A. −4915 kJ

B. −2197 kJ

C. −1789 kJ

D. −1155 kJ

25. Which set of four quantum numbers (n, l, m, s) most likely represents the valence electron in a sodium atom in its ground state?

A. $2, 1, +1, +\frac{1}{2}$

B. $3, 0, 0, +\frac{1}{2}$

C. $3, 1, -1, +\frac{1}{2}$

D. $4, 0, 0, +\frac{1}{2}$

26. If a 180 g sample of substance X had a half-life of 5 years, how many years will be needed to reduce the sample size to 11.25 g?

A. 5 years

B. 10 years

C. 15 years

D. 20 years

27. What is the mass of sulfur in 80 g of $CuSO_4$?

A. 8 g

B. 16 g

C. 32 g

D. 64 g

28. According to the following equation, 16 g of O_2 reacted with Cu_2S will produce how many grams of Cu?

$$3\ Cu_2S + 3\ O_2 \rightarrow 3\ SO_2 + 6\ Cu$$

A. 15.9 g

B. 31.75 g

C. 63.5 g

D. 127.0 g

29. What is the mass of hydrogen in 102 g of ammonia (NH_3)?

A. 2 g

B. 6 g

C. 18 g

D. 84 g

WRITTEN RESPONSE QUESTION

During cold winter months, household pipes that contain standing water sometimes burst. Explain why this occurs.

Explain why ice floats on water.

THE PRINCETON REVIEW

```
YOUR NAME: _____
(Print)         Last              First              M.I.
SIGNATURE: _____  DATE: ___/___/___

HOME ADDRESS: _____
(Print)                  Number and Street
              _____
                 City            State           Zip Code

PHONE NO.: _____
(Print)
```

Completely darken bubbles with a No. 2 pencil. If you make a mistake, be sure to erase mark completely. Erase all stray marks.

Practice Test Four

1. Ⓐ Ⓑ Ⓒ Ⓓ
2. Ⓕ Ⓖ Ⓗ Ⓙ
3. Ⓐ Ⓑ Ⓒ Ⓓ
4. Ⓕ Ⓖ Ⓗ Ⓙ
5. Ⓐ Ⓑ Ⓒ Ⓓ
6. Ⓕ Ⓖ Ⓗ Ⓙ
7. Ⓐ Ⓑ Ⓒ Ⓓ
8. Ⓕ Ⓖ Ⓗ Ⓙ
9. Ⓐ Ⓑ Ⓒ Ⓓ
10. Ⓕ Ⓖ Ⓗ Ⓙ
11. Ⓐ Ⓑ Ⓒ Ⓓ
12. Ⓕ Ⓖ Ⓗ Ⓙ
13. Ⓐ Ⓑ Ⓒ Ⓓ
14. Ⓕ Ⓖ Ⓗ Ⓙ
15. Ⓐ Ⓑ Ⓒ Ⓓ
16. Ⓕ Ⓖ Ⓗ Ⓙ
17. Ⓐ Ⓑ Ⓒ Ⓓ
18. Ⓕ Ⓖ Ⓗ Ⓙ
19. Ⓐ Ⓑ Ⓒ Ⓓ
20. Ⓕ Ⓖ Ⓗ Ⓙ
21. Ⓐ Ⓑ Ⓒ Ⓓ
22. Ⓕ Ⓖ Ⓗ Ⓙ
23. Ⓐ Ⓑ Ⓒ Ⓓ
24. Ⓕ Ⓖ Ⓗ Ⓙ
25. Ⓐ Ⓑ Ⓒ Ⓓ
26. Ⓕ Ⓖ Ⓗ Ⓙ
27. Ⓐ Ⓑ Ⓒ Ⓓ
28. Ⓕ Ⓖ Ⓗ Ⓙ
29. Ⓐ Ⓑ Ⓒ Ⓓ

Multiple-Choice Answer Key

1. D
2. B
3. A
4. B
5. C
6. A
7. B
8. A
9. C
10. C
11. C
12. A
13. A
14. C
15. B
16. D
17. B
18. A
19. B
20. D
21. A
22. A
23. A
24. C
25. B
26. D
27. B
28. C
29. C

Practice Test Four Answers and Explanations

1. **D.**

   Temperature and volume are directly proportional. As you increase temperature at a constant pressure, volume will also increase. Using Process of Elimination you can cross off all the choices except D since its volume is not greater than the original value.

   $$\frac{V_1}{T_1} = \frac{V_2}{T_2}$$

   $$V_2 = \frac{V_1 T_2}{T_1} = \frac{(50)(273)}{410} = 75 \text{ mL}$$

2. **B.**

   At 10 °C and 2 atm, the substance is a liquid as shown in the diagram below.

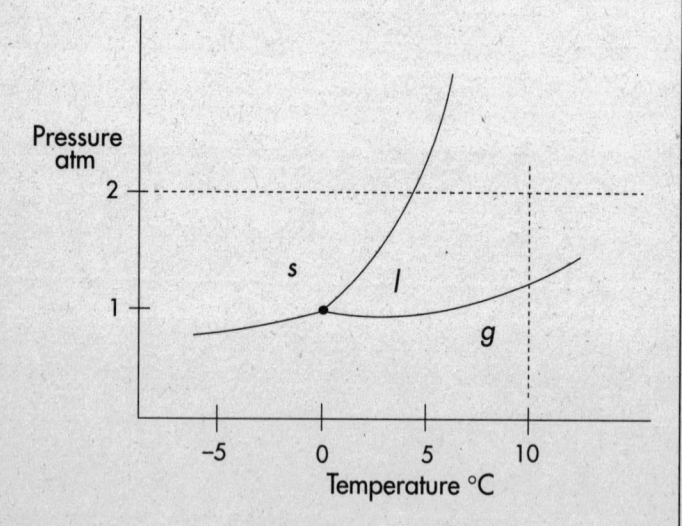

3. **A.**

   Generally speaking, electronegativity increases as you move across a period (horizontal row) on the periodic table.

4. **B.**

   First convert all terms into the necessary units:

   Temperature: 25 °C = 298 K ≈ 300 K

   Volume: 300 mL = 0.300 L

   $$PV = nRT$$

   $$n = \frac{PV}{RT} = \frac{(8)(0.3)}{(0.08)(300)} = 0.1 \text{ mol}$$

   Next convert moles of hydrogen gas to grams:

   $$0.1 \text{ mol H}_2 \left(\frac{2 \text{ g}}{1 \text{ mol}}\right) = 0.2 \text{ g H}_2$$

5. **C.**

   $$q = mc\Delta T$$

   $$\Delta T = \frac{q}{mc} = \frac{60}{(100)(0.03)} = 20 \text{ °C}$$

6. **A.**

   According to Le Chatelier's principle, a disruption in equilibrium will cause a shift in the direction of the reaction. If volume is increased, the reaction will shift towards the side that produces more moles of gas.

7. **B.**

Carbon has four valence electrons and each oxygen atom has six, resulting in a total of 16 valence electrons for a molecule of CO_2. Using a Lewis dot structure, we can distribute the 16 electrons around the central carbon atom and the bonded oxygen atoms.

$$::O::C::O::$$

This results in a double bond between the central carbon atom and each of the oxygen atoms. The two atoms will lie at opposite sides of the central atom in order to minimize the electron repulsions, resulting in a linear shape.

8. **A.**

$$\Delta H = H_{products} - H_{reactants}$$
$$= [2(CO_2) + 4(H_2O)] - [2(CH_3OH) + 3(O_2)]$$
$$= [2(-394) + 4(-242)] - [2(-201) + 3(0)]$$

Approximate
$$\approx -800 - 1000 - (-400)$$
$$\approx -1400 \text{ kJ (which is closest to choice A)}$$

9. **C.**

The polar water molecule allows hydrogen bonding to occur between a slightly positive hydrogen atom of one water molecule and a slightly negative oxygen atom of another water molecule.

10. **C.**

First, calculate the quantity in a 100.0 g sample from the percentages:

$$36\% \text{ Ca} = 36.0 \text{ g Ca}$$
$$64\% \text{ Cl} = 64.0 \text{ g Cl}$$

Next, convert grams to moles:

$$36.0 \text{ g Ca} \left(\frac{1 \text{ mol}}{40.1 \text{ g}}\right) = 0.9 \text{ mol Ca}$$

$$64.0 \text{ g Cl} \left(\frac{1 \text{ mol}}{35.5 \text{ g}}\right) = 1.8 \text{ mol Cl}$$

Next, calculate the mole-to-mole ratio between the elements:

Ratio: $\dfrac{\text{Ca}}{\text{Cl}} = \dfrac{0.9 \text{ mol Ca}}{1.8 \text{ mol Cl}} = \dfrac{1 \text{ mol Ca}}{2 \text{ mol Cl}}$

The empirical formula is therefore $CaCl_2$.

Watch out: Choice D is the correct ratio but it does not represent the simplest formula, which is the empirical formula.

11. C.

First, calculate the molecular weight of AgCl:

MW of AgCl = (108) + (35) = 143 g/mol

Next, convert grams of AgCl into moles:

$$35.75 \text{ g} \left(\frac{1 \text{ mol}}{143 \text{ g}}\right) = 0.25 \text{ mol AgCl}$$

According to the balanced equation, for every mole of AgCl produced, one mole of HCl was reacted. Therefore:

0.25 mol AgCl = 0.25 mol HCl

$$\text{Molarity} = \frac{\text{moles}}{\text{liter}}$$

The molarity of the HCl solution was then:

$$\frac{0.25 \text{ moles}}{0.05 \text{ L}} = 5 M$$

12. A.

$$P_{total} = P_{gas1} + P_{gas2}$$

$$\left(\frac{40}{100}\right) 4 \text{ atm} = 1.6 \text{ atm oxygen}$$

$$4 - 1.6 = 2.4 \text{ atm helium}$$

13. A.

First convert grams of N_2 gas into moles:

$$1.68 \text{ g } N_2 \left(\frac{1 \text{ mol}}{28 \text{ g}}\right) = 0.06 \text{ mol } N_2$$

$$PV = nRT$$

$$P_{gas1} = \frac{nRT}{V} = \frac{(0.06)(0.08)(273)}{4} = 0.33 \text{ atm}$$

14. C.

First calculate the molecular weight of NaOH.

MW = 23 + 16 + 1 = 40 g/mol

The percent by mass of Na is $\frac{23}{40}$, which is a little greater than 50% ($\frac{23}{40} > \frac{20}{40}$)

The percent by mass of O is $\frac{16}{40}$ which is 40% ($\frac{4}{10}$)

The percent by mass of H is $\frac{1}{40}$, which is a little greater than 2% ($\frac{2}{80} > \frac{2}{100}$)

Use process of elimination and you are left with C.

Watch out for answer choice A. Remember the question asks for percent composition by mass.

15. B.

Watch out! The question asks which of the following CANNOT be a nonpolar structure. This means you are looking for the one that is ALWAYS polar, no matter what atoms are bonded. A bent shape is always polar allowing for a net dipole moment.

16. D.

Moving from point A to point B as a result of decreasing pressure for water represents the change from liquid to solid, which is freezing.

17. B.

A 2.5 *m* aqueous solution is created by adding 2.5 mol of glucose to 1 kg of water. In order to calculate the mole fraction of glucose you will need to calculate moles of water.

$$MW\ H_2O = 2(1) + (16) = 18\ g/mol$$

$$1\ kg = 1000\ g\ H_2O$$

$$1000\ g \left(\frac{1\ mol}{18\ g}\right) = 55.6\ mol\ H_2O$$

The mole fraction is the

$$\frac{\text{number of moles of A}}{\text{total moles in the solution}},\ \text{so}$$

$$\frac{2.5\ \text{moles of glucose}}{2.5\ \text{moles of glucose} + 55.6\ \text{moles of } H_2O}$$

18. A.

The equilibrium constant is calculated by multiplying the concentrations of the products raised to the power equal to their coefficient, divided by the concentration of the reactants raised to the power equal to their coefficients.

19. B.

If additional K⁺ ions are being produced, then there must be a strain in the equilibrium causing the reaction to shift to the right. Look at each item separately.

I: By increasing temperature, the reaction will shift in the endothermic direction where energy is absorbed. In this reaction the endothermic direction is to the right, since energy is a reactant (on the left side of the equation). Therefore, item I is correct and you can cross off choice C.

II: By adding additional KNO_3, the reaction will also shift to the right to create K⁺ and NO_3^- ions to rebalance the concentrations. Therefore, item II is correct and you can cross off choice A.

III: By adding additional NO_3^- ions, the reaction will be shifted towards the left to counteract the additional ions. This would not cause an increase in K⁺ ions. Item III is incorrect, and the answer would be choice B.

20. D.

A solution with pH less than 7 is an acidic solution. A salt added to water would create an acidic solution if the salt was composed of the conjugate from a strong acid (Cl⁻ from HCl) and a conjugate from a weak base (NH_4^+ from NH_3). Choice A would produce a basic salt since it is composed of a conjugate from a strong base and a conjugate of a weak acid. Choice B and C are comprised of conjugates from strong acids and bases and would produce neutral solutions.

21. A.

You can use process of elimination with this question and come to the correct answer without doing any math. If we need to neutralize 50 mL of a 0.2 M NaOH solution with a solution of HCl at a greater concentration, it would require less HCl than the original 50 mL of NaOH. Therefore, cross off any choice that is not less than 50 mL. You can cross off choices B, C, and D. The correct answer is choice A.

Now let's do it the math way:

The moles of HCl and NaOH will be equal once the solution is neutralized. First calculate the moles of NaOH.

$$\text{Molarity} = \frac{\text{moles}}{\text{volume}}$$

moles = Molarity × volume = (0.2M)(50mL) = 10 mmol

Next calculate the volume of HCl that will need to be added to obtain 10 mmol of HCl.

moles = Molarity × volume = 20 mL

22. A.

Comparing experiments 1 and 2, when the concentration of [A] is held constant and the concentration of [B] is doubled, the rate of reaction stays the same. The reaction is therefore zero order with respect to [B]. (In other words, the rate of the reaction is not dependent on [B]). Comparing experiments 1 and 3, when the concentration of [B] is held constant and the concentration of [A] is doubled, the reaction also doubles. The reaction is first order with respect to [A]. The rate law is then k[A].

23. A.

Boiling point elevation is dependent on the number of particles in solution. Choice A dissociates into 3 particles, choice B does not dissociate, and choices C and D dissociate into 2 particles each. Since all of the solutions have the same molality, the greatest boiling point elevation will be the solution that dissociates into the most ions: choice A.

24. **C.**

The original equation has 4 Al on the left side of the reaction. Therefore, multiply the first equation by 2:

$$4Al(s) + 3O_2(g) \rightarrow 2Al_2O_3 =$$
$$2(-1676) = -3352 \text{ kJ}$$

Since 3 Mn are on the right side of the reaction, you need to reverse the second equation and multiply it by 3. By reversing the reaction, multiply it by −1.

$$3MnO_2 \rightarrow 3Mn(s) + 3O_2$$
$$= -3(-521)$$
$$= +1563 \text{ kJ}$$

Then add the two equations together:

$$4Al(s) + 3O_2(g) \rightarrow 3MnO_2(s)$$
$$\Delta H = +1563 \text{ kJ}$$

$$3MnO_2(s) \rightarrow 3Mn(s) + 3O_2(g)$$
$$\Delta H = -3352 \text{ kJ}$$

Therefore:

$$4Al(s) + 3MnO_2(s) \rightarrow 2Al_2O_3(s) + 3Mn(s)$$
$$\Delta H = -1789 \text{ kJ}$$

25. **B.**

Sodium's electron configuration is $1s^2 2s^2 2p^6 3s^1$. Its valence electron is located in the $3s$ subshell. Therefore $n = 3$, $l = s = 0$, $m = 0$, $s = +\dfrac{1}{2}$.

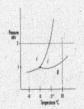

26. **D.**

Draw a chart to help you out.

Number of Years	Sample size (g)
0	180
5	90
10	45
15	22.5
20	11.25

27. **B.**

$$\text{Moles} = \frac{g}{MW}$$

$$\text{Moles of } CuSO_4 = \frac{80 \text{ g}}{160 \text{ g/mol}} = 0.5 \text{ mol}$$

Every mole of $CuSO_4$ contains 1 mole of sulfur.

$$0.5 \text{ mol S} \left(\frac{32 \text{ g}}{\text{mol}}\right) = 16 \text{ g}$$

28. **C.**

First convert grams of O_2 into moles:

$$16 \text{ g } O_2 \left(\frac{1 \text{ mol}}{32 \text{ g}} \right) = 0.5 \text{ mol } O_2$$

According to the balanced equation, every 3 moles of O_2 reacted produces 6 moles of Cu. Therefore:

$$0.5 \text{ mol } O_2 \left(\frac{6 \text{ mol Cu}}{3 \text{ mol } O_2} \right) = 1 \text{ mol Cu}$$

Finally, convert moles of Cu into grams:

$$1 \text{ mol Cu} \left(\frac{63.5 \text{ g}}{1 \text{ mol}} \right) = 63.5 \text{ g Cu}$$

29. **C.**

First calculate the molecular weight of NH_3:

$$MW = (14) + 3(1) = 17 \text{ g/mol}$$

Next, covert grams of ammonia into moles:

$$102 \text{ g } NH_3 \left(\frac{1 \text{ mol}}{17 \text{ g}} \right) = 6 \text{ mol } NH_3$$

Every mole of NH_3 contains 3 moles of hydrogen.

Therefore: $18 \text{ mol H} \left(\frac{1 \text{ g}}{1 \text{ mol}} \right) = 18 \text{ g H}$

WRITTEN RESPONSE QUESTION

This is a result of water expanding as it freezes, causing pressure on the pipes. As water freezes, hydrogen bonds are created, forming a lattice structure. This allows the molecules to be farther apart as solid ice than in water, and thus volume increases.

$$\text{Density} = \frac{\text{mass}}{\text{volume}}$$

Volume and density are inversely proportional. If volume increases, density decreases. This also allows ice to float on water.

ABOUT THE AUTHOR

Amanda Stewart has been working for The Princeton Review in Philadelphia in 1994 when she began teaching SAT, GRE, and MCAT preparation courses. She received her B.S. in chemistry from Villanova University in 1995. After graduation, Amanda moved to California and began her professional career as a project manager for a pharmaceutical company in Palo Alto. She later left the pharmaceutical industry to become the Director of Career Services for a small technical college in Hayward. She currently tutors students in chemistry and mathematics and writes test preparation books for The Princeton Review in Long Island, New York, where she lives with her husband and her dog.

NOTES

NOTES

NOTES

NOTES

NOTES

NOTES

We have a smarter way to get better grades in school.

(Go to tutor.com)

Find a tutor in 3 easy steps:

Find.
Log onto our website: **www.tutor.com**

Connect.
Sign up to find a tutor who fits all your needs

Learn.
Get **tutored** in any subject or skill

Visit www.tutor.com

FIND US...

International

Hong Kong
4/F Sun Hung Kai Centre
30 Harbour Road, Wan Chai,
Hong Kong
Tel: (011)85-2-517-3016

Japan
Fuji Building 40, 15-14
Sakuragaokacho, Shibuya Ku,
Tokyo 150, Japan
Tel: (011)81-3-3463-1343

Korea
Tae Young Bldg, 944-24,
Daechi- Dong, Kangnam-Ku
The Princeton Review- ANC
Seoul, Korea 135-280,
South Korea
Tel: (011)82-2-554-7763

Mexico City
PR Mex S De RL De Cv
Guanajuato 228 Col. Roma
06700 Mexico D.F., Mexico
Tel: 525-564-9468

Montreal
666 Sherbrooke St.
West, Suite 202
Montreal, QC H3A 1E7 Canada
Tel: (514) 499-0870

Pakistan
1 Bawa Park - 90 Upper Mall
Lahore, Pakistan
Tel: (011)92-42-571-2315

Spain
Pza. Castilla, 3 - 5° A, 28046
Madrid, Spain
Tel: (011)341-323-4212

Taiwan
155 Chung Hsiao East Road
Section 4 - 4th Floor,
Taipei R.O.C., Taiwan
Tel: (011)886-2-751-1243

Thailand
Building One, 99 Wireless Road
Bangkok, Thailand 10330
Tel: (662) 256-7080

Toronto
1240 Bay Street, Suite 300
Toronto M5R 2A7 Canada
Tel: (800) 495-7737
Tel: (716) 839-4391

Vancouver
4212 University Way NE,
Suite 204
Seattle, WA 98105
Tel: (206) 548-1100

National (U.S.)

We have over 60 offices around the U.S. and run courses in over 400 sites. For courses and locations within the U.S. call 1 (800) 2/Review and you will be routed to the nearest office.

MORE EXPERT ADVICE
from
THE PRINCETON REVIEW

Find the right school • Get in • Get help paying for it

CRACKING THE SAT & PSAT
2000 EDITION
0-375-75403-2 • $18.00

CRACKING THE SAT & PSAT WITH
SAMPLE TESTS ON CD-ROM
2000 EDITION
0-375-75404-0 • $29.95

THE SCHOLARSHIP ADVISOR
2000 EDITION
0-375-75468-7 • $25.00

SAT MATH WORKOUT
0-679-75363-X • $15.00

SAT VERBAL WORKOUT
0-679-75302- • $16.00

CRACKING THE ACT WITH
SAMPLE TESTS ON CD-ROM
2000-2001 EDITION
0-375-75501-2 • $29.95

CRACKING THE ACT
2000-2001 EDITION
0-375-75500-4 • $18.00

CRASH COURSE FOR THE SAT
10 Easy Steps to Higher Score
0-375-75324-9 • $9.95

DOLLARS & SENSE FOR COLLEGE
STUDENTS
How Not to Run Out of Money by
Midterms
0-375-75206-4 • $10.95

PAYING FOR COLLEGE WITHOUT
GOING BROKE, 2000 EDITION
Insider Strategies to Maximize Financial
Aid and Minimize College Costs
0-375-75467-9 • $18.00

BEST 331 COLLEGES
2000 EDITION
The Buyer's Guide to College
0-375-75411-3 • $20.00

THE COMPLETE BOOK OF COLLEGES
2000 EDITION
0-375-75462-8 • $26.95

THE GUIDE TO PERFORMING ARTS
PROGRAMS
Profiles of Over 600 Colleges, High
Schools and Summer Programs
0-375-75095-9 • $24.95

POCKET GUIDE TO COLLEGES
2000 EDITION
0-375-75416-4 • $9.95

AFRICAN AMERICAN STUDENT'S GUIDE
TO COLLEGE
Making the Most of College: Getting In,
Staying In, and Graduating
0-679-77878-0 • $17.95

With Free Apply! Software

WE ALSO HAVE BOOKS TO HELP YOU SCORE HIGH ON
THE SAT II, AP, AND CLEP EXAMS:

CRACKING THE AP BIOLOGY EXAM 2000-2001 EDITION
0-375-75495-4 • $17.00

CRACKING THE AP CALCULUS EXAM AB & BC 2000-2001 EDITION
0-375-75499-7 • $18.00

CRACKING THE AP CHEMISTRY EXAM 2000-2001 EDITION
0-375-75497-0 • $17.00

CRACKING THE AP ECONOMICS EXAM (MACRO & MICRO) 2000-2001 EDITION
0-375-75507-1 • $17.00

CRACKING THE AP ENGLISH LITERATURE EXAM 2000-2001 EDITION
0-375-75493-8 • $17.00

CRACKING THE AP U.S. GOVERNMENT AND POLITICS EXAM 2000-2001 EDITION
0-375-75496-2 • $17.00

CRACKING THE AP U.S. HISTORY EXAM 2000-2001 EDITION
0-375-75494-6 • $17.00

CRACKING THE AP PHYSICS 2000-2001 EDITION
0-375-75492-X • $19.00

CRACKING THE AP PSYCHOLOGY 2000-2001 EDITION
0-375-75480-6 • $17.00

CRACKING THE AP EUROPEAN HISTORY 2000-2001 EDITION
0-375-75498-9 • $17.00

CRACKING THE AP SPANISH 2000-2001 EDITION
0-75401-4 • $17.00

CRACKING THE CLEP 4TH EDITION
0-375-76151-9 • $20.00

CRACKING THE SAT II: BIOLOGY SUBJECT TEST 1999-2000 EDITION
0-375-75297-8 • $17.00

CRACKING THE SAT II: CHEMISTRY SUBJECT TEST 1999-2000 EDITION
0-375-75298-6 • $17.00

CRACKING THE SAT II: ENGLISH SUBJECT TEST 1999-2000 EDITION
0-375-75295-1 • $17.00

CRACKING THE SAT II: FRENCH SUBJECT TEST 1999-2000 EDITION
0-375-75299-4 • $17.00

CRACKING THE SAT II: HISTORY SUBJECT TEST 1999-2000 EDITION
0-375-75300-1 • $17.00

CRACKING THE SAT II: MATH SUBJECT TEST 1999-2000 EDITION
0-375-75296-X • $17.00

CRACKING THE SAT II: PHYSICS SUBJECT TEST 1999-2000 EDITION
0-375-75302-8 • $17.00

CRACKING THE SAT II: SPANISH SUBJECT TEST 1999-2000 EDITION
0-375-75301-X • $17.00

THE PRINCETON REVIEW
Visit Your Local Bookstore or Order Direct by Calling 1-800-733-3000
www.randomhouse.com/princetonreview

www.review.com

Expert Advice

Talk About It

www.review.com

Pop Surveys

Paying for it

www.review.com

THE PRINCETON REVIEW

Getting in

Word du Jour

www.review.com

Find-O-Rama School & Career Search

www.review.com

Best Schools

Finding it

www.review.com